DON'T PANIC: HERE'S THE TRUTH ABOUT AI

SHED THE HYPE. CATCH YOUR BREATH. CHART THE COURSE.

GLEN DAY

PROJECTOR HOUSE

DON'T PANIC

HERE'S THE TRUTH ABOUT AI

CONTENTS

ACKNOWLEDGMENTS

I want to acknowledge my family and friends. You give my life meaning beyond and deeper than the technology life I experience through screens. I certainly need to keep my head in the real world more than I do. Thank you for keeping me at least somewhat tethered to reality.

Thank you also to the constant flow of people who have joined me in random conversations on airplanes, in coffee shops, at conferences, along hiking trails, and in those unforced human moments of sharing that happen before online meetings come to order. Those conversations are fuel for life that carry over beyond the constant flow of missions.

And a big thanks to the late Douglas Adams, who inspired the title of this book. In *The Hitchhiker's Guide to the Galaxy*, he noted that the Guide had "outsold the *Encyclopedia Galactica* and become the standard repository of knowledge for two main reasons: it is slightly cheaper, and it has the words 'DON'T PANIC' inscribed in large, friendly letters on its cover."

You know, Adams was onto something.

We can get through this, whether it's AI, self-driving cars, or something more along the lines of a Vogon Constructor Fleet. Just don't panic.

INTRODUCTION

Author's note

The world feels like it is spinning entirely too fast. Like it's going to fling us off. You feel it, right? It's not just me?

Let's stir some actual intelligence in with the artificial.

Every time I look at my phone, the ground rules of reality seem to have shifted. A new AI model drops. A new trillion-dollar valuation is announced. A tech billionaire predicts a utopian age of limitless leisure, while a competing pundit warns of an impending, jobless apocalypse.

If you are feeling a low-grade hum of existential vertigo, I'm with you. We are experiencing the collective whiplash of a society that has just been handed fire for the second time, and we are all still arguing over whether it's going to cook our dinner or burn down the house.

Let's get one thing straight right out of the gate: The shift we are living through is real. This isn't a passing fad. It isn't a Meta-verse simulation. It isn't a crypto-bro fever dream. The arrival of reasoning AI is a tectonic event on par with the printing press, the steam engine, and the internet. It is going to fundamentally rewire how we work, how we learn, how we create, and how we interact with the sum total of human knowledge. We are watching the marginal cost of digital intelligence drop to zero, and that is going to change the world in profound, irreversible ways.

And yes, it is going to be messy.

We are going to face real, structural challenges. We are going to stress-test our global power grids and argue over the environmental cost of server farms. We are going to navigate a tsunami of synthetic content, deepfakes, and hallucinations that will force us to question our own eyes and ears. We are going to have deeply uncomfortable, society-wide negotiations about what human labor is actually worth when a machine can draft a legal brief, write code, or conceptualize a brand campaign in three seconds.

There are smart people out there pointing to real challenges. We should listen and learn. While Yuval Harari's books on the future are brilliantly researched and probably aren't wrong in the observations that have earned Harari the nickname Dr. Doom, perhaps it's worth having something in your life that gives you not just fear but also a sense that we're not all on a predestined collision course with hopelessness. And it's good for you to know that I'm not just telling you that and putting DON'T PANIC in big, friendly letters on the cover to sell books. I'm doing so because the story doesn't end with change. Something else comes after that, and we'll still be us—for better or worse.

So, momentarily set down the Harari, switch off the cable news, and step away from the social feed... even if just for a bit. We should hear out folks like Harari. And then we should move along with building our communities, loving our families and neighbors, and making this as good a life as we can.

We can do this.

There is a difference between a *disruption* and an *apocalypse*.

The panic you are feeling right now is the natural human response to a sudden loss of control. But as we will explore in this book, that fear is being actively weaponized. You and I are being sold a narrative of doom by an ecosystem of media outlets, consultants, and algorithms that literally profit off our elevated heart rates.

When we step away from the breathless, hyper-reactive echo chambers of Silicon Valley—a place and mindset I work with and need to step away from daily—it is much easier to see the forest for the trees. The real world does not move at the speed of a tech demo. In reality, it moves at the speed of human friction and our own stubborn human nature.

This book is your deep breath. It is a permission slip to step off the ledge.

In the following pages, we are going to systematically dismantle the panic. We will take the most terrifying headlines of our era—the vanishing jobs, the algorithmic bosses, the death of human art, and the end of shared truth—and move them into the clear light of reality. We are going to separate the marketing hype from the actual mechanics of the machine.

You don't need a PhD in computer science to survive the next decade, and you certainly don't need to retreat to an off-grid cabin. You just need context. Panic is a reaction to the unknown. Calm is the byproduct of understanding.

By the end of this journey, you won't just understand what AI is capable of. You will understand what it *isn't*. You will see that the machine isn't here to replace the human soul—it is here to remove the friction that gets in its way. You don't need to outrun the Engine. You just need to learn how to drive it.

––––––––

The Collective Whiplash

It is hard to pinpoint the exact Tuesday it happened, but the whiplash was universal.

One minute, Artificial Intelligence was a fun, distant sci-fi concept. It was a neat parlor trick that could generate weird pictures of astronauts riding horses, or a distant supercomputer that could beat a grandmaster at chess. It was something for Silicon Valley engineers to tinker with while the rest of us went to work.

The next minute, it was writing your kid's history essay, passing the bar exam, and quietly threatening to do your job better, faster, and for a fraction of the cost.

If you are feeling a low-grade, constant thrum of anxiety right now, you are not alone. You are just paying attention. Every time you open your phone, there is a new headline perfectly engineered to spike your cortisol. We are told that algorithms are coming for our creative skills and our paychecks. We are warned that deepfakes are erasing the line between truth and fiction, and that massive data centers are going to drain the power grid dry. We have tech billionaires and researchers issuing apocalyptic warnings about the end of human usefulness.

It feels like we are all trapped in the backseat of a car that has suddenly accelerated to a hundred miles an hour, and there is no one at the steering wheel. (Of course, all things considered, that analogy itself feels quite a bit less dramatic than it would have just ten years back.)

The panic we are feeling isn't just about the technology itself. It is about a profound, disorienting loss of control. It is the suffocating feeling that the fundamental rules of the world were rewritten while we were sleeping, and if we don't figure out how to speak to the machine by tomorrow morning, we are going to be permanently left behind.

———

Separating Science from Fiction

But here is the good news: you don't need a computer science degree to take the wheel. You don't need to learn how to code, and you certainly don't need to start prepping for a digital apocalypse.

What you actually need right now is a translator.

As a creative and content director, my job has often been just that—to translate the possibilities of tomorrow. Working with global agencies, I've crafted the narratives for tech titans and automotive innovators—companies like Google, AWS, Microsoft, Qualcomm, and Audi. I've had a front-row seat in the rooms where AI systems and autonomous vehicles are imagined and marketed. At the same time, serving on university advisory boards for AI and multimedia design has shown me the other side of the equation: how the shockwaves of this massive disruption are hitting the students and professionals whose futures will be most directly shaped by these breakthroughs.

That vantage point has taught me one crucial, liberating lesson: there is a massive difference between the marketing hype engineered to sell software, services, or subscriptions and the physical, practical reality of what these machines can actually do.

The tech industry thrives on framing its innovations as unstoppable forces of nature. A tidal wave is intimidating. It makes you feel like a helpless bystander who just has to brace for impact and hope you don't drown in the disruption. But AI isn't the weather. It is infrastructure. It is a highly sophisticated, incredibly fast, but fundamentally steerable tool.

The promise of this book is simple. We are going to look under the hood and demystify the machine. We are going to strip away the complex jargon, the corporate buzzwords, and the sci-fi doom-scrolling. The goal here is not to turn you into a prompt engineer or a Silicon Valley insider. The goal is to give you the clarity and the context to stop competing with artificial intelligence and start directing it.

———

The Roadmap

To learn how to steer this new road, we first have to clear the fog. That is exactly what we are going to do in the pages ahead.

This book is designed as a guide, divided into two distinct parts.

In the first part, we are going to walk straight into the dark. We cannot build a solid foundation until we systematically dismantle the fears that are keeping us awake at night. We will confront the identity crisis of the white-collar worker and the artist. We will look at the exhausting realities of the algorithmic boss, the entirely justified anxieties over our stolen data, the physical limits of our power grids, and the looming sci-fi dread of a superintelligent takeover. We aren't going to casually dismiss these fears. We are going to unpack them, understand the realities behind them, and ultimately disarm them.

Then, once the panic has subsided and the noise has cleared, we will pivot.

In the second half, we will open **The Architect's Playbook**. This is your practical, everyday manual for moving from defense to offense. You don't need to rebuild your entire life to survive the AI era. You just need a new set of rules. We will break down the five core pillars of living alongside the machine. You will learn how to shift from being a "Doer" to a "Director," how to stay continuously relevant by becoming a "Five-Minute Student," and how to aggressively delegate your digital drudgery so you can buy back your time.

Most importantly, we will explore why your "Proof of Human"—your messy, physical, analog reality—is about to become your single most valuable asset, both at work and at home.

———

The Deep Breath

We have been standing at this exact precipice before. Every time a civilization-altering technology arrives, it feels like the end of the world for the people living through the transition.

When the printing press was invented, critics warned it

would destroy human memory and unravel the fabric of society. When the locomotive arrived, people literally feared that traveling at thirty miles an hour would cause the human body to disintegrate. When the internet connected the globe, we were terrified it would replace physical communities entirely. Disruption is always uncomfortable, and it always displaces the old ways of doing things. But it is also the very soil where our next great leap forward grows.

Artificial intelligence is the latest, and undoubtedly the fastest, shift we have ever faced. But it is still just a tool in a very long lineage of human inventions.

The future is not a predetermined script written by a server farm in Silicon Valley. It is not a tidal wave that is simply happening *to* you. It is a landscape, and you have the agency to decide how you want to build on it. You get to decide which tasks you hand over to the machine, and which parts of your messy, beautiful humanity you fiercely protect.

So, before we step into the chaos and confront the fears of the next chapter, I want you to do one thing.

Take a breath.

And, don't panic.

You're about to know the truth about AI.

Now, about that career you've worked so hard for year after year…

CHAPTER 1

OH, NO! AI IS TAKING MY JOB?

The fear of mass unemployment versus the reality of task automation and shifting roles.

It happens at dinner parties, in private Slack channels, and at 2:00 AM when you are staring at the ceiling. It's the creeping, cold-sweat realization that the digital tool you used to summarize a meeting this morning might eventually decide it no longer needs you to attend the meeting at all. Or perhaps the fear is that your boss—the one who talks about how the whole company is one big family, then lays off a bunch of said family every year—realizes that some new AI does your job better than you.

For the first time since the Industrial Revolution, the anxiety of automation has breached the gates of the white-collar world. While it may be bringing newly sleepless nights to knowledge workers, factory workers and warehouse packers have lived with this specter for decades. But now, it's the copywriters, the financial analysts, the junior coders, and the corporate strategists who are feeling the floorboards rot beneath their feet.

If you are feeling this existential dread, you are not necessarily being paranoid. You are doing something seemingly smart: paying attention.

The psychological toll of this shift is measurable and severe. According to a landmark Gallup poll, the percentage of college-educated workers who fear their jobs will be made obsolete by technology more than doubled in a remarkably short window following the mainstream release of generative AI. The American Psychological Association now actively tracks technology and AI-driven displacement as a significant, compounding source of national workplace stress. We are experiencing a collective, society-wide bout of professional vertigo.

But before you abandon your career and look into off-grid farming, we need to take a hard look at *why* you are so terrified. What exactly have you been paying attention to?

Panic at this scale rarely happens by accident. It's always sparked by something. And sometimes it's manufactured.

While there will certainly be disruption to the workforce—some fields have been seeing it for years already—there's something else contributing to the anxiety. The narrative that AI is a mechanical grim reaper coming for your livelihood is being actively stoked by voices with a massive vested interest in keeping your heart rate up.

- **The Media Ecosystem:** News outlets do not generate clicks with headlines that say, "AI Will Gradually Shift Your Daily Workflows." They generate revenue with catastrophic warnings of economic collapse. Fear is the ultimate algorithm hack.
- **The Consulting Industrial Complex:** Massive advisory firms need to sell multi-million-dollar "AI Readiness" and "Digital Transformation" rescue packages to terrified CEOs. To sell the expensive cure, they first have to convince the corporate world that it has a terminal disease.

- **Dystopian Authors and Pundits:** There is a booming cottage industry of tech-philosophers who build their personal brands by positioning themselves as the lone prophets of an impending doom. And frankly, it's a no-lose game plan. If you keep predicting the end of the world, you'll eventually be right.
- **Click-bait Kings:** It's not all big business. These days, a single social influencer can pump out hundreds of posts a day and reach millions, or even a Mr. Beast-like billions of audience members. And whether those clicks come from dopamine or cortisol, they pay the same.

To build this narrative of mass unemployment, these industries rely on a specific tactic: **Weaponized Data**.

They take highly nuanced, deeply researched macroeconomic reports from the world's most boring financial institutions, strip them of all their stabilizing context, and feed them to you as proof of the apocalypse.

To cure the panic, we have to look at the actual numbers they are using to scare you. Let's unpack the three biggest "Headline Horrors" fueling the panic today, starting at 300 million.

———

The Goldman 300

If you've opened up LinkedIn or scrolled a tech blog lately, you've very likely seen a headline like: ***"Goldman Sachs warns AI will destroy 300 million jobs."***

Content creators, doomsday preppers, and click-hungry journalists have used this specific number as the ultimate proof that the AI job-eating monster is upon us. It's presented as a literal body count—300 million people handed pink slips, packing up their desks, and being permanently replaced by a chatbot. It feeds directly into the anxiety that we are all about to become obsolete.

The Hype: Goldman Sachs predicts doom

This one line came from a 31-page macroeconomic research paper and grew into an economy-eating monster.

The report said that generative AI could "expose the **equivalent** of 300 million full-time jobs to automation". That was rewritten as "300 million jobs **lost**."

Those are two profoundly different things. If an AI automates 25% of the daily administrative tasks for four different people, that mathematically equals "one full-time job equivalent" of automated work. Does it mean someone gets fired? Not necessarily. It can mean their time was put toward more pressing tasks —and there are plenty of them in this era of scrambling to figure out the new opportunities and beat everyone to them. The people who quoted the report swapped the word "tasks" for "humans" because fear generates clicks and clicks mean advertising dollars.

The Reality: Goldman predicts opportunity

If you read the actual report (published in Spring 2023 by economists Joseph Briggs and Devesh Kodnani), you'll find it isn't a doomsday manifesto at all. It is overwhelmingly optimistic. Here is what the panic-mongers intentionally leave out:

- **Complemented, Not Substituted:** The report explicitly states that for the vast majority of workers, AI will *complement* their work, not replace it. In the U.S., they estimated only about 7% of jobs are at risk of total substitution, while 63% will be complemented by AI, and 30% will be completely unaffected.

- **The Productivity Boom:** Far from destroying the economy, Goldman Sachs predicted that this automation will trigger a massive global boom. They estimate generative AI could actually increase global US GDP by 7%. That translates to adding nearly $7 trillion to the US economy over ten years.
- **The Reinvestment of Time:** The economists noted that when workers have their "freed-up capacity" (because the AI is handling the busywork), they naturally reallocate that time toward more productive, higher-value activities.

The biggest thing the doom-scrollers miss about the Goldman Sachs report is its heavy emphasis on historical precedent.

The authors point out that while technology has *always* displaced certain tasks, it consistently creates entirely new industries and wealth to replace them. The report cites a staggering economic statistic to prove this: **60% of today's workers are in occupations that didn't even exist in 1940.**

This implies that over 85% of employment growth in the last 80 years was driven directly by the technology-driven creation of *new* positions. Before the internet, there was no such thing as a "Web Designer" or "Cloud Architect." The report argues that AI will do the exact same thing—spawning millions of new roles we don't even have the names for yet.

The Bottom Line: The Goldman Sachs report wasn't predicting a jobless dystopia. It was predicting a massive, highly profitable upgrade to human output. The "300 million" number isn't the number of people who will be unemployed. It's the sheer volume of *busywork* that is about to be wiped off our collective plates.

———

The McKinsey Body Count

If the Goldman Sachs report provided the "body count," **McKinsey & Company** provided the "timeline of doom."

The Hype: McKinsey says millions will be on the streets

When people hear that the $500/hr consultants at McKinsey predict **"30% of work hours could be automated by 2030"** and **"375 million people will be forced to shift careers,"** the panic meter spikes like an Olympic volleyball team taking on your middle school gym class. The narrative online and in breakrooms is that within a few short years, a third of the economy will simply evaporate, and hundreds of millions of people will be left wandering the streets with useless degrees, desperately trying to reinvent themselves in a world that no longer needs them. It sounds less like a technological shift and more like a mass extinction event for the working class.

Once again, the fear machine took a complex macroeconomic forecast and stripped it of all its stabilizing context. They didn't say **"30% of work hours could be automated by 2030"** and **"375 million people will be forced to shift careers"**.

The folks using this report to up the click rate conflated **"work hours"** with **"headcount."** If an AI automates 30% of your daily workload by doing things like summarizing meetings, drafting emails, or organizing spreadsheets, you don't get 30% fired.

Second, the phrase **"completely shift careers"** is weaponized to mean "become destitute." The hyped headlines paint these 375 million occupational transitions as sudden, tragic layoffs. They ignore that "shifting careers" is a constant, historically normal feature of a healthy, evolving economy.

The Reality: McKinsey notes need

When you dig into McKinsey's actual research, including their landmark updates on Generative AI and the Future of Work, the picture isn't a dystopian collapse; it's a massive, necessary labor reorganization. Here is what the panic-merchants leave out:

- **The Productivity Imperative:** McKinsey explicitly states that the global economy *needs* this automation. With aging populations and shrinking birth rates in the developed world, we literally do not have enough human workers to maintain global economic growth over the next two decades. We *need* the AI to pick up the 30% slack just to keep our living standards from dropping.
- **The "Unbundling" of the Job:** The report stresses that very few occupations will be automated out of existence entirely. In reality, it's usually estimated at less than 5%. Instead, jobs will be "unbundled." The boring, repetitive tasks will be handed to the machine, leaving the human worker to focus on the tasks that require high-level strategy, empathy, and physical dexterity.
- **The Trillion-Dollar Upside:** McKinsey estimates that Generative AI alone could add the equivalent of $2.6 trillion to $4.4 trillion annually to the global economy. That is a whole lot of new wealth that will be pumped directly into creating new industries, new demands, and entirely new job categories.

The fundamental blind spot in the way people are reporting the McKinsey data is the assumption that the economy is a "fixed pie." People assume that if a robot does a job, there is one less job for a human. Economics doesn't work that way.

What the reporters ignore is that **automation lowers the cost of goods and services, which increases demand.** If AI allows a software company to build an app in half the time, they don't necessarily fire half their coders. While it might mean the cost of off-the-shelf software plummeted, it may also mean there's new opportunity for every small business to afford custom software. If that happens, demand explodes, and the company actually ends up hiring *more* coders to manage the AI outputs and meet the new demand. And if the company doesn't represent this opportunity, then the one that does, may well hire away their coders before they're ever laid off.

To that point, McKinsey points out that the "career shifts" are largely driven by massive new investments in other areas. We are simultaneously going through an energy transition that requires millions of green energy workers and a demographic shift that requires millions of healthcare workers. The workforce isn't being thrown away; it's being redeployed to where humans are actually needed.

The Bottom Line: The McKinsey report isn't a warning that you will be replaced; it is a forecast that your daily routine is about to be violently upgraded. The 375 million people transitioning aren't victims of a robot takeover. They are the vanguard of a workforce moving away from drudgery and toward higher-value human skills.

The World Economic Displacement

When the **World Economic Forum** releases a report stating that **"83 to 92 million jobs will be displaced,"** the collective internet gasps. Social media influencers, tech pessimists, and forum threads, on the other hand, salivate. Their outlets collectively explode with the same terrifying narrative: a workforce roughly the size of Germany is about to be wiped off the map.

The anxiety here is deeply structural. People aren't just afraid for their own specific roles; they are terrified of a societal collapse where nearly a hundred million people are rendered permanently unemployable by a massive, unavoidable wave of automation.

When the World Economic Forum releases a report stating that "83 to 92 million jobs will be displaced," the collective internet gasps. Social media influencers, tech pessimists, and forum threads explode with the same terrifying narrative: a workforce roughly the size of Germany is about to be wiped off the map.

The anxiety here is deeply structural. People aren't just afraid for their own specific roles; they are terrified of a societal collapse where nearly a hundred million people are rendered permanently unemployable by a massive, unavoidable wave of automation.

The Hype: WEF predicts the end

The misquoters take ample liberties with the WEF's complex analysis of global workforce churn projection and turn it into a terrifying subtraction problem. **The WEF didn't say "83 to 92 million jobs will be displaced" and just leave it at that.**

The fear-feeders latch onto the word "displaced" and present it as "destroyed" or "permanently unemployed." Their headlines take "83 million jobs" from the WEF's 2023 report and "92 million jobs" from their 2025 projections without ever mentioning the second half of the very same sentence. They frame "displacement" as a cliff you fall off, rather than a bridge you cross to a new role.

The Reality: WEF predicts growing demand

If you actually read the WEF's *Future of Jobs* reports, you find that the narrative of mass destruction is a mathematical illusion. Here is what the panic-merchants leave out:

- **The "Creation" Side of the Equation:** The media focuses exclusively on the displaced jobs, but conveniently ignores the newly created ones. The 2025 WEF report explicitly projects that while 92 million roles face displacement, **170 million new jobs will be created**. That isn't a loss. That is a massive net gain of 78 million jobs globally.
- **The Shift Away from Routine:** The jobs slated for displacement are highly routine, repetitive data-processing and clerical roles (like data entry or payroll clerks). In contrast, the demand for analytical thinking, complex problem-solving, and creative roles is skyrocketing.
- **The Boom Beyond Tech:** The report notes that not all of this churn is even about AI. Massive job creation is coming from the green energy transition, agriculture, and growth in the care economy driven by aging populations. Furthermore, the expected automation of physical and manual work has actually *decreased* in recent projections.

The real warning in the WEF report isn't a lack of jobs. It's a lack of **updated skills**.

What the doom-scrollers miss is the WEF's heavy emphasis on "structural churn"—the idea that the global economy is rapidly trading old skills for new ones. The panic assumes that if a data-entry clerk's job is automated, that human being simply ceases to participate in the economy.

The WEF argues the exact opposite. They say 85% of surveyed employers plan to prioritize internal upskilling to bridge this gap. The goal is to retrain that clerk to manage the new data systems, or to help them shift into booming sectors like education or healthcare where human empathy and dexterity are at a premium.

The Bottom Line: The WEF isn't predicting a jobless future; it is predicting a *different* future. The 83 to 92 million "displaced" jobs represent the shedding of outdated routines to make room for 170 million new opportunities. According to the WEF, the robot isn't taking your livelihood; it's forcing a massive, global shift in what it means to be a human worker.

The point here is not that nothing is going to happen or that some types of jobs won't go away. Huge changes are in the works and many types of jobs will either evolve greatly or be replaced by others, but that's not exactly a new pattern. This is not the panic situation many are unscrupulously reporting it to be for their own gain. It is a situation that will offer both challenge and opportunity on a grand scale.

Of course, it's not just the media, bloggers, and social accounts out there trying to wind you up. They're not the only ones who either don't bother to proof their claims or actively don't care that they are bending the truth. There are whole rows of bookstores. Playing the same game for the same reason— getting a rise out of you pays.

———

The Dystopian Bookshelf

Walk into any major bookstore, head to the Business or Technology section, and look at the covers. You'll notice a distinct, almost comical aesthetic. Black backgrounds. Neon red or glowing green fonts. Ominous silhouettes. Titles that lean heavily on words like *End, Death, Useless, Apocalypse,* or *Jobless.*

We need to make a careful distinction here: There are brilliant, well-researched books written by serious economists and technologists who harbor genuine, well-founded concerns about the friction of the AI transition. We should absolutely read and debate those.

But that's not what is driving the mass anxiety. We are talking about the **Doomsday Industrial Complex**—the authors and self-proclaimed "futurists" who willfully lean into the fear because they know that panic guarantees a spot on the bestseller list.

The story these books sell is always absolute. The machine won't just *change* your job. It will *eradicate* your industry overnight. Humans will be reduced to obsolete pets living on government handouts.

These authors take isolated tech demos—an AI passing a medical exam or writing a passable sitcom script—and draw a straight, uninterrupted line to total human economic collapse. They sell you the idea that the future is already written, and you are not in it.

These books aren't actually about the future of technology. They're about the economics of the publishing industry. Nuance doesn't get you booked on morning talk shows or secure a massive speaking fee. "Things will change, but we will adapt," doesn't sell a million hardcovers.

Doing this research and writing from my desk in Colorado, far outside the frantic Silicon Valley echo chamber, provides a necessary detachment. Out here, you remember that the real world doesn't move at the speed of a sensationalist tech-thriller. These doomsday books intentionally ignore the most powerful force in the global economy: human friction. They assume that regulations, labor unions, corporate bureaucracy, consumer preferences, and basic human stubbornness will simply evaporate. They ignore the fact that even if a machine *can* do something perfectly today, it takes years—often decades—for society to actually reorganize around it, leaving plenty of time for new human roles to emerge.

The Bottom Line: A lot of what is sitting in the non-fiction aisle is actually science fiction. The authors of these panic-primers aren't predicting your future; they are monetizing your anxiety. If a book tells you with 100% certainty that your career is over, put it back on the shelf. The future is a negotiation, not a fore-closure.

———

Doom Sites

It would be one thing if this manufactured panic stayed confined to clickbait articles and the business section of the bookstore. But the fear is no longer theoretical. It has breached the psychological hull, and it is actively changing human behavior right now. Namaste, everyone.

Right now, based on the projections we're hearing, we are making massive, life-altering decisions—shifting our education, abandoning our career tracks, and giving up on our ambitions—all based on the illusion that the game is already over.

Let's look at a few examples of how this anxiety is being spread and manifested in the real world, and inject a little reality back into the room.

The Doomsday Clock

In early 2026, a free online tool called *TheGreatDisplacement.ai* went viral. The premise was morbidly simple: type in your exact job title, and the algorithm spits out a personalized "displace-ment year"—the exact date an AI will allegedly render you obsolete.

Tools like this turn abstract economic anxiety into a personal, ticking countdown clock. It feels authoritative because it is a computer telling you when a computer will replace you. FWIW, I tried it, myself. The title of the job I came up with in advertising, Copywriter, shot up on the color scale right past the friendly greens and benign yellows into the threatening oranges (not quite all the way to terrifying reds you get if you type in "data entry"). It listed my job automation risk at 70% with an esti-mated year of 2028. It wasn't specific about what automation meant to my job. It didn't say explicitly that I'd be out of a job.

But shock of all shocks, it did have a message reading "Want a step-by-step plan to AI-proof your career? SmartOwner sends you one free lesson every day." right next to a big red button screaming "SUBSCRIBE FREE".

A couple of things. First, I entered my other job description, Creative Director—which it wrongly equated with "film/video director"—I was in the friendly green spectrum with only a 20% automation risk. So, let's just say calculators like this have seriously questionable veracity. Second, we all know "SUBSCRIBE FREE" doesn't mean that the landing page we arrive at is going to be a public service page made from the goodness of someone's heart, don't we. Let me tell you as a longtime copywriter with apparently high odds of being replaced by AI in the next two years, who has written the words on such buttons more than a few times, "SUBSCRIBE FREE" really means, "BUY THIS."

These calculators are the digital equivalent of a carnival fortune teller. They operate on highly static, linear assumptions. They look at what a job is *today* and assume the job will never evolve. They ignore the fact that as AI takes over the repetitive tasks, the human has new opportunity to move up the value chain to manage the output.

The Bottom Line: Don't let a parlor trick built on flawed assumptions dictate your blood pressure. Your career doesn't have an expiration date. It has an evolution date.

The College Major Pivot

The panic is trickling down to the next generation. Gen Z students, terrified by the headlines, are actively shifting their college majors. They are abandoning fields like Computer Science, corporate administration, and the liberal arts in favor of highly physical, "robot-proof" roles like Nursing or the trades. They believe white-collar salaries are about to evaporate entirely.

Sitting on university advisory boards including strategic AI programs, I see a heartbreaking irony to this trend. Students may be fleeing the exact fields that are about to experience the greatest leverage in human history. We still desperately need coders, strategists, and creatives—but we need them to direct the AI, not do the manual typing.

When you work with AI, you come to realize its job is to tell you what you want to hear. As a result, we can't have just anyone prompting our AI. A person with no insight into what successful code, marketing, philosophy, writing, etc. looks like, is not going to know whether or not AI is putting out something of value. We need more than just plumbers and HVAC professionals in our future. As much as ever, we will need people with critical insights looking for specific valuable answers and recognizing the difference. Fleeing knowledge work out of fear could actually mean abandoning the greatest productivity boom of the 21st century.

The Bottom Line: The world will absolutely need nurses and electricians, but they won't be the only people left with jobs. Running toward physical labor solely because you are afraid of a chatbot could be a massive strategic error. The knowledge workers of tomorrow won't be typists. They will be conductors.

CHAPTER 2

THE IDENTITY CRISIS

For generations, we have introduced ourselves to strangers with a single, defining question: *"What do you do?"* It's a simple greeting, but it reveals a deeper truth about how we view ourselves. In the modern world, our identity is inextricably linked to our output. We are the sum of our skills. We are the crafts we spent decades practicing, the degrees we went into debt to earn, and the professional titles we fought to secure. We have built our entire sense of self-worth around our ability to be useful, productive, and uniquely capable.

Then, almost overnight, the rules of the game changed.

For decades, we were comforted by a very specific narrative about automation: the robots were coming, but they were only coming for the manual labor. Machines were supposed to lift heavy boxes, assemble engines, and sweep floors. They were meant to relieve us of the physical grind so that human beings could elevate themselves to the higher pursuits of the mind. Algorithms were never supposed to paint watercolors, write heartfelt poetry, formulate business strategies, or compose music.

But the script has flipped. The machines bypassed the physical world and went straight for the cognitive one.

We are now facing a profound psychological shock. This isn't just the economic fear of losing a paycheck; it's the existential terror of losing our purpose. When a piece of software can instantly replicate the specialized knowledge you spent ten thousand hours mastering, the crisis goes far deeper than your bank account. It strikes at the core of who you are.

In this chapter, we are going to explore the front lines of this collision between human identity and artificial intelligence. We will unpack:

- **The Death of the Artist:** The emotional toll of watching algorithmic perfection commoditize the human soul and the creative process.
- **The Speed Trap:** The exhausting, accelerating pace of technological change that makes us feel like we are constantly falling behind.
- **The Expertise Evaporation & The Skills Cliff:** The terrifying reality that traditional mastery and hard-earned knowledge may no longer guarantee a safe harbor.
- **The Algorithmic Boss & The Resume Black Hole:** The maddening, Kafkaesque experience of being managed, judged, and ultimately rejected by unfeeling lines of code.
- **From the Breadline to the Blurred Line:** How the historical anxiety of the factory floor has officially migrated to the laptop screen.

This is the great white-collar reckoning. If we are no longer defined solely by what we can produce, we have to answer a much harder question: what exactly is our value in the marketplace?

Let's step into the void, starting with the very people who thought they were the safest of all.

———

The Death of the Artist

If the institutional economic reports provided the "body count," generative AI tools that created images like Nano Banana and writing like ChatGPT provided the existential crisis.

When people see a machine win a state fair art competition, or watch an algorithm spit out a highly stylized brand logo and a 500-word blog post in three seconds, the panic meter spikes. The narrative online and in creative Slack channels is that within a few short years, the commercial arts will simply evaporate.

The Hype: No one needs an artist

The fear is that copywriters, graphic designers, illustrators, and authors will be left wandering the streets with useless portfolios, desperately trying to reinvent themselves in a world where "good enough" is suddenly free. It sounds less like a technological shift and more like the end of the human soul in commerce.

Once again, the fear machine took a massive technological leap and stripped it of all its practical, day-to-day context. They looked at a machine that could mimic style and declared that human creativity was officially dead.

The folks using this narrative to up the click rate equated "generating content" with "creative problem-solving." It's not the same. Take advertising agencies, for example. If an AI can instantly generate fifty headlines or mock up a dozen campaign concepts, do you fire your creative team and replace them with a marketing AI like Jasper? Not if you want your agency to have a reason for existence. Creativity is the product of an ad agency. That's true whether it's the development of a creative headline, a creative media approach, or a creative new idea of ways to connect a client's message with their audience.

The clients already have their own AI that can write headlines and make cool images. If an agency lets go of its creativity, the clients will soon let go of the agency. This is true of creativity in all kinds of aspects. It is not the ability to string together clever words or draw impressive pictures. It is the ability to relate and express in new and better ways. AI doesn't replace that ability—it amplifies it.

Second, the phrase "AI-generated art" is weaponized to mean "the end of human value." The hyped headlines paint these new tools as sudden, tragic replacements for human ingenuity. They ignore that the tools of creation have constantly evolved—from the paintbrush to the camera to Photoshop—and each time, the human creator simply adapted to a higher level of abstraction.

The Reality: There's a creative revolution coming

When you dig into the actual reality of working in the trenches of a commercial ad agency or staring down the blank page of a new book manuscript, the picture isn't a dystopian collapse of creativity. It's a massive, necessary reorganization of how we ideate. Here is what the panic-merchants leave out:

- **The Premium on Taste:** AI models are, at their core, prediction engines. They are trained on the past, meaning they generate the mathematical average of what has already been done. They can produce perfectly competent, grammatically correct, visually balanced work. But in the creative world, whether that be an ad agency, concert hall, theater, gallery, film set, stage, Etsy account, or any other variation, "average" doesn't sell. It takes a human director's *taste* to look at fifty AI-generated concepts, throw out forty-nine of them, and pick the one remaining idea that has those countless key elements needed to actually capture human attention.

- **The "Unbundling" of the Blank Page:** Very few creative occupations will be automated out of existence entirely. Instead, the creative process is being "unbundled." The most tedious parts of the job—staring at a blank page, building rough storyboards, writing placeholder copy, or doing basic background research—can be sped up by AI. This leaves the human worker to focus on the tasks that require high-level strategic alignment, emotional resonance, and brand voice.

- **The Authenticity Upside:** When synthetic content becomes free and infinite, the basic laws of supply and demand kick in. The market is about to be flooded with cheap, AI-generated noise. In that environment, verifiable, human-crafted work complete with its unique quirks, lived experiences, and original perspectives becomes a premium. Originality, creativity, taste, vision: these traits become even more valuable and distinct because they're framed by a world overflowing with generic, mass-produced pieces lacking, quite literally, heart.

The fundamental blind spot in the way people are reporting the "Death of the Artist" is the assumption that the creative economy is a "fixed pie." People assume that if a robot writes a paragraph or generates an image, there is one less paragraph or image for a human to make. Economics doesn't work that way.

What the panic-drivers miss—or ignore—is that automation lowers the cost of *ideation*, which increases the demand for *great execution*. If AI allows a writer to explore more aspects of a book, or a filmmaker to explore more potential scenes or edits, that doesn't mean fewer of those artists are needed, but rather it means the quality of the output can rise. The same goes for an ad agency, an architecture firm, or any other business that trades on the quality of its creative output. Just because the creative teams can ideate faster doesn't mean fewer creative minds are needed. It means the market standard of success is going to rise and creative professionals who know how to use the emerging AI tools to raise the quality of their output will be in higher demand.

The Bottom Line: The rise of generative AI isn't a warning that the artist will be replaced; it is a forecast that their daily routine is about to be upgraded. The creatives transitioning into this new era aren't victims of a robot takeover; they are the vanguard of a workforce moving away from the drudgery of the blank page and toward the much higher-value skill of human taste.

———

The Speed Trap

There is a hidden danger in this new era of instantaneous generation, and it isn't that the machine is going to steal the final product. It's also not the kind folks are panicking about. The real danger is that the machine is going to steal the *friction*.

The Hype: AI will generate creativity at speed

Any working creative—whether they are arranging a multi-track audio session, dialing in the brand voice for a massive tech campaign, or outlining a non-fiction book—will tell you a funda-mental truth about the job: **The execution *is part of* the ideation.** It is incredibly rare to sit down with a perfectly formed, crys-talline idea, execute it flawlessly from start to finish, and end up with exactly what you pictured in your head. That is how a printer works, not a human mind.

The Reality: Creativity will always take time

Instead, true creativity requires layers of discovery. You write a scene, and halfway through, you realize the supporting character is actually the hero. You spend three days cutting together a promotional video, make a mistake on the timeline, and realize the accidental jump-cut is the exact pacing the spot needed. You start building a brand strategy, get stuck in the mud of the competitive research, and are forced to pivot to an entirely new, vastly superior angle.

That pivot—the "happy accident"—only happens because the human brain is allowed to wander during the slow, methodical process of *doing the work*. One of the most defining characteristics of talent is the ability to recognize the gift in the occasions of serendipity. And it's a loss if they go away.

When generative AI threatens to collapse that timeline from three weeks to three seconds, this vital and usually unappreciated friction can be lost. When you type a prompt and receive a finished deliverable instantly, you bypass the wandering. You get exactly what you asked for, which is often the most dangerous thing that can happen to a creative project. The first idea is rarely the best idea; it's just the most obvious one.

Engineering Intentional Slowness

To survive the Zero-Marginal-Cost era of infinite content, we cannot simply let the machine sprint from brief to final draft. If we lose the evolutionary time of the project, the work will become perfectly competent and perfectly soulless.

Instead, the creators of tomorrow must learn how to artificially introduce "slow time" back into the process, using the AI's speed to buy back their own capacity to wander.

Here is what that counterbalance looks like in practice:

- **The Rapid Wrong Turn:** Instead of using AI to quickly generate the *final* answer, use it to instantly generate fifty *wrong* answers. Spend the first hour of a project commanding the AI to visualize every obvious, cliché, and terrible direction the campaign could possibly take. Get the bad ideas out of the system at the speed of light.
- **The Breadth of the Sandbox:** AI gives you the power of infinite variation. If you lose the time you used to spend iterating on one idea, counterbalance it by exploring ten radically different conceptual universes before lunch. You are shifting the "discovery phase" from the *middle* of the project to the very *beginning*.
- **Intentional Friction (The Human Override):** Once the rapid ideation phase identifies the brilliant, weird thread worth pulling, take the machine's hands off the wheel. Create boundaries where the AI is no longer allowed to drive. Slow down. Write the final copy yourself. Tinker with the audio mix manually. You aren't doing this because the AI *can't* finish it; you are doing it because the microscopic decisions made during the physical act of creation are what embed the genius hidden in the firing of our trillions of synapses into the work.

The Bottom Line: The machine can accelerate the production, but it cannot automate the epiphany. If we allow AI to handle both the conception and the final execution in a single keystroke, we strip the art of its humanity. The true masters of this new era will be the ones who know exactly when to let the algorithm sprint, and when to force the whole process to a grinding, highly intentional halt.

———

The Expertise Evaporation

Writers, artists, and data entry clerks aren't the only people nervous about job loss. There is a constant flow of posts warning of the impending doom to knowledge workers when AI reaches par with humans in every level of thought. This level of AI is popularly referred to as Artificial General Intelligence, and predictions of its arrival range from tomorrow to 30 years from now to never.

If generative AI coming for the artists was an existential crisis, its arrival in the realms of law, medicine, and engineering feels like an economic earthquake.

For decades, the white-collar middle-class dream was built on a very specific, reliable formula: take out massive student loans, spend years in graduate school memorizing a vast amount of complex, highly regulated information, and then charge clients by the hour to access the database inside your head. You were an "Information Gatekeeper."

But then the headlines hit: *AI Passes the Bar Exam in the 90th Percentile. AI Diagnoses Breast Cancer Better Than Top Radiologists. AI Writes Code Faster Than Senior Developers.* Suddenly, the foundation of the professional class seemed to crack.

The Hype: The "Napster-ing" of the Professional Class

The panic here is absolute: Why would a client ever pay a lawyer $500 an hour to draft a contract when an AI agent can do it in four seconds for fractions of a penny? Why wait three weeks to see a specialist if an app can diagnose your rash with perfect accuracy right now?

The narrative peddled in the doom-scroll is that the "Zero Marginal Cost" of intelligence will completely hollow out the middle class. The fear is that once high-end expertise becomes a free digital commodity, the millions of professionals who make up the backbone of the economy will be rendered entirely useless, replaced by server farms and subscription fees.

The Reality: The Illusion of the "Information" Economy

This panic is built on a fundamental misunderstanding of what we are actually buying when we hire a professional. We *think* we are buying information. We aren't. We are buying **accountability, judgment, and risk mitigation.**

When you dig into the day-to-day reality of high-stakes professions, the narrative of "expertise evaporation" completely falls apart. Here is what the panic-merchants aren't saying:

- **The Accountability Premium:** An AI can draft a brilliant, airtight commercial lease. But if that lease contains a hallucinated clause and your business gets sued, you cannot depose an algorithm. You cannot sue a chatbot for malpractice. Corporations and individuals will always pay a premium for a human to review the machine's work, sign their name to the bottom line, and say, *"I stake my professional reputation and my legal liability on this."* We don't just pay professionals to *know* things. We pay them to take the blame if things go wrong.
- **The Empathy Mandate:** In healthcare, diagnosis is only about 20% of the job. The other 80% is care. An AI might be able to read an MRI with superhuman precision, but an AI cannot sit in a sterile room, look a terrified patient in the eye, and help them navigate the emotional devastation of a cancer diagnosis. AI will undoubtedly handle the data-crunching of medicine, which will finally free up doctors and nurses to do the one thing the machine can't: actually spend time healing the human being in front of them.

- **The Explosion of Demand (Jevons Paradox):** In economics, there is a concept called Jevons Paradox, which states that when technology makes a resource cheaper and more efficient to use, demand for that resource tends to *increase*. Right now, millions of small businesses operate without legal counsel because it's too expensive. Millions of people put off medical care because the system is too bloated. If AI makes delivering these services 80% cheaper and faster, the market doesn't shrink—it explodes. Lawyers won't be fired; they will suddenly be able to take on five times as many clients, serving a massive population that was previously priced out of the market.

We are not watching the death of expertise. We are watching the death of *rote memorization*. The machine is taking over the role of the encyclopedia, which means the human is finally free to be the strategist.

The Bottom Line: The rise of AI isn't a warning that the professional middle class will be wiped out; it is a forecast that the nature of their value is shifting. The doctors, lawyers, and engineers who thrive in the next decade won't be the ones who can memorize the most facts. They will be the ones who demonstrate the best judgment, the highest emotional intelligence, and the willingness to take responsibility for the machine's output. The expertise isn't evaporating; it's just moving from the head to the gut.

———

The Skills Cliff

If the "Expertise Evaporation" is the fear that clients won't need you, the "Skills Cliff" is the deeply personal terror that you won't even know how to need yourself.

It's the quiet anxiety that creeps in when you use an AI to summarize a 40-page PDF, write a python script, or draft a sensitive client email. You look at the polished result and realize: *I didn't actually do the work. If the Wi-Fi goes down tomorrow, could I still do this job from scratch?*

For a lot of professionals, the answer is starting to feel like a terrifying "No."

The Hype: The "WALL-E" Brain

The narrative peddled by tech-pessimists is one of rapid cognitive atrophy. They warn that we are outsourcing our brains to the cloud. The fear is that if we let the machine do the heavy lifting of thinking, structuring, and drafting, a whole generation of junior workers will never learn the fundamentals of their craft.

The doom-scroll paints a picture of a hollowed-out workforce —a generation of "prompt engineers" who don't actually know how to write, code, or strategize, sitting helplessly at their desks the moment the algorithm throws an error. It's the fear that we are walking right off a cognitive cliff, happily giving away the mental friction that actually makes us smart.

The Reality: The Evolution of Cognitive Load

This panic is based on a massive historical amnesia. We have been having this exact same panic attack for centuries, every single time a new technology offloads a piece of human memory or mechanical effort.

When the pocket calculator was introduced, math teachers warned that students would forget how to do arithmetic and the nation's mathematical capability would plummet. When GPS arrived, critics warned we were losing our spatial awareness and our ability to read paper maps.

Did we lose those specific, mechanical skills? Yes. Did we become dumber? No. We simply shifted our cognitive load. Here is what the panic-merchants are missing about the way humans actually work:

- **From Drafting to Directing:** In the trenches of building a brand strategy or directing a creative project, the most valuable skill isn't the physical act of typing out the brief or painstakingly formatting the slides. The value is in the *critique*. By letting the AI generate the messy first draft, junior workers are essentially being forced to immediately act like senior directors. They aren't learning how to type. They're learning how to *edit*, how to evaluate logic, and how to spot a weak argument. They are training their taste.
- **The "Black Box" of Fundamentals:** The fear assumes that unless you do the agonizing manual labor, you don't understand the underlying concepts. But you don't need to know how to hand-write assembly code to be a brilliant software engineer today, just like you don't need to know how to manufacture a canvas to be a master painter. AI is simply abstracting the next layer of fundamentals, allowing human brains to operate at a higher, more strategic altitude.
- **The Bandwidth Boom:** The human brain has a finite amount of working memory. When you spend three hours wrestling with the syntax of a spreadsheet formula or the grammar of a report, you have less mental bandwidth available for the actual *insight* that data reveals. By outsourcing the rote mechanics of the job to the AI, we are freeing up massive amounts of neurological bandwidth to focus on complex problem-solving and interpersonal connection.

The "Skills Cliff" is an illusion created by looking backward. We aren't losing our skills; we are trading them in for better ones. We traded the ability to read a Thomas Guide for the ability to navigate anywhere on Earth instantly. We are now trading the ability to write a mediocre first draft for the ability to curate a brilliant final product.

The Bottom Line: The machine isn't making us forget how to work; it is forcing us to redefine what "work" actually is. You aren't falling off a cognitive cliff. You are stepping onto an escalator. The professionals who embrace this won't lose their minds—they will become the editors and directors of a much larger, faster digital workforce.

———

The Algorithmic Boss

If the "Skills Cliff" is the fear that you will forget how to do your job, the "Algorithmic Boss" is the fear of who exactly will be grading you on it.

It is the dystopian dread of logging into your computer and realizing your daily performance, your annual bonus, and your ultimate employment status are no longer being decided by a human being. Instead, you are being evaluated by an emotionless optimization engine that tracks your keystrokes, measures your eye movements on Zoom, and penalizes you for spending too long staring out the window.

They Hype: The "Robo-Manager"

The panic here is fueled by very real, terrifying stories from the gig economy and logistics sectors. We've all read the articles about delivery drivers or warehouse workers being automatically terminated by an app because they missed a delivery window or their "time off task" metric ticked a percentage point too high.

The doom-scroll narrative takes this blue-collar reality and projects it onto the entire white-collar world. The fear is that the empathy of a human manager—the boss who understands that you missed a deadline because your kid had a fever, or that you were staring at the ceiling because you were thinking through a complex problem—is about to be replaced by a cold, calculating dashboard. The anxiety is that we will all be reduced to a "productivity score" generated by a machine that doesn't understand what it means to be human.

The Reality: The "Context Deficit"

This panic is built on a fundamental misunderstanding of what a manager in the knowledge economy actually does. An AI is incredible at tracking data, but it is entirely incapable of understanding *context*.

When you dig into the reality of directing teams and managing talent, the narrative of the all-powerful "Robo-Boss" hits a massive wall. Here is what the panic-merchants are missing:

- **The Limits of Quantification:** In a warehouse, output is easily measured by volume and speed. But in the knowledge economy, value is not tied to keystrokes. In my day-to-day work directing creative and content teams, I can tell you firsthand that a strategist might spend three days reading and thinking, producing only a single, brilliant paragraph. An algorithm tracking keystrokes would flag them for termination. A human manager gives them a promotion. AI cannot measure the quality of an epiphany.
- **The Legal and HR Firewall:** Corporate America is fundamentally risk-averse. Allowing a black-box algorithm to automatically fire, demote, or evaluate employees based on synthetic metrics is a legal

nightmare. It opens the company up to massive liabilities regarding algorithmic bias and discrimination. Human Resources departments and legal teams will mandate that a "Human-in-the-Loop" must always have the final say on employment decisions. The machine can flag an issue, but only a human can pull the trigger.

- **The "Copilot" for Leadership:** Just as AI is unbundling the tasks of the worker, it is unbundling the tasks of the manager. Right now, most middle managers spend 80% of their time acting like human routers—updating spreadsheets, tracking project statuses, and allocating resources. AI is going to completely automate that administrative friction. This won't eliminate the manager; it will finally free them up to do the actual work of *leadership*: mentoring, coaching, navigating fragile egos, and building team culture.

We are not watching the death of the empathetic boss. We are watching the death of the *administrative micromanager*. The machine is taking over the role of the clipboard, which means the human boss is finally free to be a leader.

The Bottom Line: The AI isn't your new boss; it is your boss's new assistant. The companies that thrive won't be the ones that use algorithms to ruthlessly squeeze their employees for an extra ounce of productivity. They will be the ones that use AI to eliminate the bureaucratic friction, giving human managers the time and bandwidth to actually lead their people.

———

The Resume Black Hole

If the "Algorithmic Boss" is the fear of how you will be managed, the "Resume Black Hole" is the deeply demoralizing fear of how you even get your foot in the door in the first place.

We have all felt it: you spend three hours meticulously tweaking your bullet points to match a job description, hit "Submit" on a corporate portal, and receive an automated rejection email exactly 14 seconds later. It feels less like a job hunt and more like shouting into an uncaring digital void.

The Hype: The "Digital Trash Can"

The doom-scroll tells us that corporate HR departments have handed the keys to our livelihoods entirely over to biased AI gatekeepers. The panic is that these algorithms are ruthlessly scanning resumes for exact keyword matches, silently throwing 90% of applicants into the digital trash before a human eye ever glances at the page.

The fear is twofold: First, that if you don't know the exact "SEO cheat codes" for your resume, you are invisible. Second, that these systems are baking in historical prejudices, silently locking perfectly qualified people out of the economy based on cold, algorithmic math.

The Reality: The End of the Keyword Game

Here is the truth: getting your resume ignored by a computer isn't a new AI problem. We have been dealing with "dumb" Applicant Tracking Systems (ATS) for twenty years. Those old systems *did* require exact keyword matches. But the new era of generative AI is actually fixing the black hole, not deepening it.

When you look at how hiring managers and creative directors are actually adapting to this tech, the narrative of the all-powerful AI gatekeeper falls apart. Here is what the panic-merchants are missing:

- **Context Over Keywords:** Old software would auto-reject you if the job asked for a "Content Manager" and your resume said "Editorial Director." It was a rigid, stupid matching game. Modern AI language models actually understand *meaning*. They know those two titles require the exact same core competencies. Ironically, the new AI gatekeepers are vastly superior at recognizing untraditional backgrounds, career pivots, and transferable skills than the old software ever was.

- **The Two-Way Arms Race:** The AI isn't just on the employer's side. It's on yours. Candidates are using AI to instantly tailor their resumes and write flawless cover letters for every single application. But here is the catch: because *everyone* now has a perfectly optimized, AI-assisted resume, the resume itself is losing its power as a filter. It is becoming a basic entry ticket, not the deciding factor.
- **The Legal Firewall:** Corporations are terrified of algorithmic bias lawsuits. Laws are already on the books mandating independent bias audits for AI hiring tools. HR departments aren't blindly trusting the machine; they are treating it like a highly suspicious intern. A "Human-in-the-Loop" is still required to make the final call, because while a machine can verify your work history, it cannot assess cultural fit, adaptability, or sheer human drive.

We are not watching the death of the fair application process. We are watching the death of the resume as the ultimate arbiter of talent.

The Bottom Line: The AI isn't throwing your application in the trash. It's simply raising the baseline. In a world where every applicant can instantly generate a flawless, machine-optimized PDF, the piece of paper matters less. Your actual human proof—your portfolio, your network, your reputation, and your taste—matters more. You aren't fighting a robot gatekeeper; you are finally being freed to prove your value beyond a bulleted list.

———

From the Breadline to the Blurred Line

Take a deep breath. The robots are not marching on your office.

The 300 million Goldman Sachs "body count" was a mathematical illusion. The algorithmic boss has a legal leash. The blank page isn't dead; it has just been upgraded. As we've seen, the Zero-Marginal-Cost era isn't going to hollow out the human workforce. It is going to elevate it. We are trading the drudgery of the rough draft for the high-value work of human judgment, empathy, and taste.

Whether you are directing international ad campaigns for tech giants, managing a local storefront, or simply trying to navigate your own career pivot, the core lesson remains the same: you get to keep your job. You are no longer just the producer; you are the director.

But here is the catch. As you step into this new role as the "Curator of the Machine," you are about to face a completely different kind of vertigo.

If the AI is doing the executing—generating text, code, images, and data at the speed of light—the fundamental nature of our digital world is about to change. The machine doesn't just flood the market with cheap work. It floods the market with cheap *reality*.

When you ask an AI to summarize a report, how do you know it didn't just invent a statistic that sounded highly plausible? When you see a video of a political candidate, or get a frantic voice memo from a family member, or review a perfectly formatted portfolio from a new job applicant, how do you know it's actually real?

The panic is shifting. The anxiety of the late 2020s isn't

economic; it is epistemological. We are moving from the fear of obsolescence to the fear of deception.

The greatest threat to your career—and your sanity—in the next decade isn't that you will be replaced by a machine. It is that you will confidently make a massive, life-altering decision based on a machine's very convincing lie.

You don't need to panic about the unemployment line anymore. But as we turn the page to Chapter 2, you might want to start worrying about how to trust your own eyes.

Welcome to the Truth & Reality Panic.

CHAPTER 3

THE TRUTH & REALITY PANIC

Who can we trust when seeing is no longer believing?

Imagine your phone rings at 3:00 PM on a Tuesday. The caller ID is a local number. You answer, and it's your kid. They are crying, frantic, saying they've been in a minor accident and immediately need you to wire a few hundred dollars to a towing company so the police don't get involved. The voice is unmistakably theirs—the exact cadence, the slight hesitation when they are stressed, the familiar pitch. Your heart rate spikes, your adrenaline surges, and you reach for your wallet.

Except your kid is actually sitting safely in their middle school math class.

What you just heard was a cloned voice, generated by a scammer who scraped a three-second clip of your child speaking on a public TikTok video. It cost the scammer less than a dollar to make, and it took about twenty minutes.

We have spent our entire lives operating on a very simple, biological baseline of trust: *Seeing is believing. Hearing is verifying.* If someone said something on camera, or if you heard their voice on the phone, it happened. But as we cross into the late 2020s, that fundamental rule of human interaction is breaking down.

When the marginal cost of generating text, audio, and video drops to zero, it isn't just marketing copy and software code that floods the market. It is deception. We are entering an era of infinite, highly personalized, zero-cost friction. And the panic it is causing is entirely justified.

As a society, we are suddenly grappling with a terrifying new set of questions. When you ask a machine for a critical fact, how do you know it isn't just hallucinating a highly plausible lie? When a political video goes viral the night before an election, how do you know it isn't a synthetic deepfake designed to crash the polls? When an AI can perfectly mimic the style of your favorite author or artist, is it innovation, or is it just the greatest intellectual property heist in human history?

The anxiety here isn't about the economy. It is at the root of knowledge: *epistemological,* if you want the fancy term. It is the creeping dread that we are moving into a "Post-Truth" world where our personalized algorithmic feeds will completely destroy our shared social fabric, leaving us isolated in our own synthetic realities.

But before you throw your smartphone into the nearest river and disconnect your router, take a breath.

We have survived "the end of truth" before.

In the 1990s, when Photoshop became widely available, the cultural panic was identical. Pundits warned that photographs could never be trusted again, that the justice system would collapse because visual evidence was dead, and that society would fracture. But the sky didn't fall. We simply adapted. We developed digital forensics. We learned to look for the "seams" in the pixels. We evolved a new, collective digital literacy.

We are about to do the exact same thing with the Engine.

Yes, the threat is real, and the transition will be chaotic. But you don't need to be a victim of it. To turn down the panic, you need to understand exactly how the deception works, where its limits are, and how you can protect yourself and your family.

In this chapter, we are going to dismantle the "Reality Panic" piece by piece. We will look at:

- **The Deepfake Dilemma:** The terror of forged video and audio.
- **The Hallucination Hazard:** Why "brilliant" machines lie with absolute confidence.
- **The Plagiarism Machine:** The moral anxiety of AI and intellectual property.
- **The Scam Supercharger:** How to protect your bank account from personalized fraud.
- **The End of Shared Reality:** Will algorithms destroy our political common ground?

You are not powerless in this new world. You just need to upgrade your spam filter from your inbox to your brain.

—————

The Deepfake Dilemma

Imagine you log into a routine, mid-morning video call.

On the screen is your company's Chief Financial Officer, along with a few other familiar faces from leadership. The CFO, looking straight into the camera and sounding exactly like they do in every quarterly all-hands meeting, explains that the company is executing a highly confidential, time-sensitive acquisition. They ask you, as a trusted member of the finance team, to authorize an immediate wire transfer to a new vendor. You see their facial expressions. You hear the specific cadence of their voice. You nod, hang up, and execute the transfer.

Except the CFO wasn't on the call. Neither was anyone else.

The entire meeting was synthetic. The faces and voices were

deepfakes generated in real time by a scammer who scraped public corporate videos and built digital puppets.

This isn't a sci-fi movie pitch; this exact scenario actually happened to a multinational finance worker in early 2024, resulting in a $25 million loss. And it represents the visceral, stomach-dropping core of the deepfake dilemma.

The Hype: The End of Truth

The doom-scroll tells us that we are entering a "Media Apocalypse." The panic is that because an AI can generate a pixel-perfect video of a politician declaring war, or perfectly clone a CEO's voice to authorize a fraudulent wire transfer, society is going to collapse into complete, paralyzing paranoia.

The fear isn't just that we will be fooled by the fakes. It is also the threat of the "Liar's Dividend"—the terrifying reality that guilty people will simply start waving away real, damning evidence (like a leaked audio recording) by confidently claiming, "It wasn't me, it was just an AI deepfake." The narrative sold to the public is that our shared visual and auditory reality is permanently broken.

The Reality: The Shift from "Detection" to "Provenance"

This panic assumes that our defense mechanisms are going to stand perfectly still while the offensive technology accelerates. But that isn't how the tech sector, the legal system, or human psychology works.

When advanced photo editing arrived in the 1990s, pundits declared the absolute death of photographic evidence. We didn't solve the Photoshop problem by banning the software or abandoning photography. We solved it by developing a new baseline of visual literacy. We learned to look for the digital seams.

With generative AI, we are going through that exact same growing pain, but on a larger, faster scale. Here is what the panic-merchants are missing about how society is actually adapting:

- **The "Nutrition Label" for Content:** The tech industry is quietly shifting its focus from the impossible task of *detecting what is fake* to the highly solvable task of *proving what is real*. Cryptographic initiatives like the C2PA (Coalition for Content Provenance and Authenticity) are rapidly becoming a global standard. Think of it like a digital chain of custody. In the near future, cameras, browsers, and social feeds will automatically display the verifiable origin of a video.

If a sensational news clip doesn't carry the cryptographic "receipts" proving it was captured by a real camera in real time, your brain will be trained to treat it like a cheap tabloid rumor.

- **The Arms Race of Watermarking:** The digital immune system is adapting in real-time. Major AI developers are embedding imperceptible digital watermarks directly into the pixels and audio frequencies of AI-generated content. These signals survive compression and editing, allowing platforms to automatically flag or restrict synthetic media before it ever reaches your feed.
- **The Human Adaptation:** Right now, deepfakes are shocking because the novelty hasn't worn off. But human skepticism is a highly adaptive, evolutionary trait. Within a few short years, our baseline assumption when seeing an outrageous, unverified video online will naturally shift from *"This must be real"* to *"This is probably synthetic."* We won't lose our minds. We will simply raise our standards for evidence.

The Bottom Line: The deepfake dilemma isn't going to destroy the truth; it is just going to make the truth require a receipt. You don't need to live in a state of constant, exhausting paranoia. You simply need to upgrade your mental spam filter, demand verifiable sources before you share a piece of outrage, and remember that just because a machine *can* fake reality doesn't mean society is going to let it go unchecked.

———

The Hallucination Hazard

Imagine you hire a highly recommended, incredibly articulate research assistant. You ask them to find a legal precedent or a specific medical study to include in a massive presentation. An hour later, they hand you a beautifully formatted, highly convincing three-page summary of a 2022 federal court ruling. You present it to the board. The board stares at you in stunned silence.

The ruling doesn't exist. The judge doesn't exist. Your assistant simply made the whole thing up because it sounded like exactly what you wanted to hear.

This is the Hallucination Hazard. And it is the reason why so many professionals are terrified to actually integrate AI into their high-stakes workflows.

The Hype: The Broken Oracle

The doom-scroll loves a public AI failure. We have all seen the headlines about the New York lawyers who were sanctioned by a judge for submitting a legal brief filled with fake case law generated by ChatGPT. We have seen medical chatbots invent entirely new, non-existent side effects for common drugs.

The panic here is that these systems are fundamentally unreliable and dangerously deceptive. The fear is that because the machine speaks with absolute, unwavering confidence, we are going to make catastrophic professional, financial, or medical decisions based on beautifully written garbage. The narrative is that if you can't trust the AI 100% of the time, you can't trust it *any* of the time.

The Reality: The Eager Intern, Not the Encyclopedia

This panic stems from a profound public misunderstanding of what a Large Language Model (LLM) actually is.

Google has conditioned us to treat the blinking cursor like a search bar. We think we are querying a database that retrieves facts. We aren't. Generative AI is not a database. It's a prediction engine. It does not "know" what is true or false. It calculates the statistical probability of what the next word in a sentence should be, based on billions of pages of human writing.

When I advise business leaders and students on strategic AI implementation, the very first thing we have to break is the "Oracle Myth." The AI is not an infallible god of knowledge. It is a brilliant, lightning-fast, pathological people-pleaser. If it doesn't have the exact fact you asked for, it doesn't always want to disappoint you by saying, "I don't know." Instead, it predicts what a correct answer *would* look like, and hands you a hallucination.

But once you understand the mechanics of the hallucination, you can easily engineer the risk right out of your workflow. Here is what the panic-merchants ignore about how we actually manage the machine:

- **The Rise of RAG (Retrieval-Augmented Generation):**
  The tech industry is already solving the hallucination problem through a process called Grounding or RAG. Instead of asking the AI to pull facts out of its hazy mathematical memory, we are giving it the facts first.

You upload your company's actual financial PDF, or your specific legal contract, and you command the AI: *"Answer my question using ONLY the information in this document."* When you restrict its sandbox to your verified data, the hallucination rate plummets to near zero.

- **The "Trust but Verify" Tax:** We need to apply the exact same standard to AI that we apply to human beings. If a junior copywriter hands you a statistic about the B2B tech sector, you ask for the source link before you publish it. If the AI gives you a fact, you do the exact same thing. You don't fire the copywriter; you just check their work. AI speeds up the drafting process by 90%, leaving you plenty of time to spend the remaining 10% fact-checking.

- **Knowing the Right Job for the Tool:** You don't use a hammer to drive a screw. Generative AI is a *reasoning* and *ideation* engine. It is perfect for brainstorming, summarizing, translating, and drafting. It is currently the wrong tool for blind, unverified factual recall.

The Bottom Line: The machine isn't actively trying to deceive you. It's just calculating probabilities. You don't need to panic about bad data if you stop treating the AI like a magical encyclopedia. Treat it like a brilliant, over-eager intern: give it clear boundaries, provide it with the correct source material, and never publish its work without putting your own human eyes on the final product.

———

The Plagiarism Machine

As an author and someone who has spent years directing creative teams, I know exactly what it feels like to stare down a blank page. The anxiety is real, but so is the profound sense of ownership when you finally crack the spine of a new idea.

So when you watch a generative AI spit out a flawlessly formatted 500-word essay, or generate an image perfectly mirroring the style of a working illustrator, the reaction is deeply visceral. It doesn't just feel like a neat technological trick; it feels like a violation.

The Hype: The Greatest Heist in History

The narrative dominating the doom-scroll right now is that AI is nothing more than a giant, soulless copy-paste machine.

The panic is that every output is just a digital Frankenstein's monster, stitched together from the stolen, uncompensated labor of millions of human artists, journalists, and authors. We see the massive, multi-billion-dollar lawsuits—*The New York Times* suing OpenAI, record labels suing AI music generators—and we assume the entire foundation of the technology is built on a crime.

The fear here isn't just about the tech companies. It's personal. If you use AI to brainstorm a marketing campaign or outline a chapter, the anxiety is that you are actively participating in this heist, rendering your own work morally bankrupt and legally actionable.

The Reality: The Difference Between "Reading" and "Regurgitating"

To defuse this panic, we have to look past the sensational headlines and understand how the machine actually "learns," and how the legal and economic systems are currently adapting to it.

When you sit in the room with the engineers building these tools, you quickly realize that the "copy-paste" narrative is a fundamental misunderstanding of the technology. Here is what the panic-merchants are leaving out:

- **The Concept of the "Weight":** When an AI model is trained on a million books, it does not store those books on a hard drive to be retrieved later. It reads them, learns the mathematical relationship between the words, and then discards the original files. It saves the *pattern*, not the *prose*. If a human author reads a hundred sci-fi novels and then writes their own space opera based on the tropes they learned, we call it inspiration. The courts are currently wrestling with whether it is fair to call it "theft" when a machine does the exact same thing.
- **The Shift to Licensing:** The era of the "wild west" data scrape is already ending. Tech companies are reading the room (and the lawsuits). In recent years, we have seen massive, multi-million-dollar licensing deals signed between AI developers and major publishers like *The Associated Press, News Corp*, and *The Atlantic*. The market is actively correcting itself, moving from a model of unauthorized scraping to a model of paid, licensed data partnerships.
- **Plagiarism is a Human Choice:** We need to separate *copyright infringement* (a legal issue regarding how the model was trained) from *plagiarism* (a moral issue regarding how you use the output). The machine cannot plagiarize; only a human can. If you use AI to generate fifty terrible ideas to help you find one good one, that is a tool. If you use AI to write an entire chapter, paste it into a document, and claim you wrote every word, that is a moral failure on *your* part, not the algorithm's.

The fundamental blind spot in the "Plagiarism Machine"

panic is the assumption that human creativity is just the sum total of our vocabulary. It isn't.

The Bottom Line: Generative AI is a brilliant synthesizer, but it has no lived experience. It has never had its heart broken, never felt the sting of a failed project, and never sat up at 2:00 AM worrying about the future. The machine isn't here to steal your voice. It's here to force you to lean into the authentic, messy, deeply human perspectives that an algorithm can never replicate.

———

The Scam Supercharger

For the last twenty years, we had a very reliable, built-in defense mechanism against digital fraud: scammers were mostly terrible writers.

We all knew the signs. If an email from "Bank of Amerrica" started with "Dear Valued Costumer" and demanded an urgent wire transfer to a foreign prince, you rolled your eyes and hit delete. The grammar was the tell. The friction of the scam was high, and the success rate was low.

But generative AI just completely eliminated the grammar tell.

The Hype: The Golden Age of Fraud

The panic in the doom-scroll is that AI has handed the keys to the kingdom directly to cybercriminals. The fear is that we are entering an era of "polymorphic," hyper-personalized spear-phishing.

Instead of a generic, misspelled email, the scammer uses an AI agent to scrape your LinkedIn, read your public social media posts, and instantly draft a flawless, contextually accurate email that references a real project you are working on, mimicking the exact tone of your boss.

Worse is the terror of the "Family Emergency Scam." The anxiety is that an algorithm can scrape a few seconds of your elderly parent's voice from a Facebook video, clone it perfectly, and call you at midnight claiming they are in the hospital and need an immediate digital transfer. The narrative is that our bank accounts, our businesses, and our vulnerable family members are completely defenseless against a machine that can perfectly mimic the people we trust most.

The Reality: Friction is the Antidote

The threat here is absolutely real. The speed and scale at which AI can generate deceptive content is staggering. But the panic assumes that because the *hook* has become high-tech, our *defense* must also require a computer science degree.

It doesn't. When you strip away the technological novelty of a voice clone or a flawless AI email, the core anatomy of a scam hasn't changed at all. Scammers still rely entirely on two deeply human emotions: **Urgency and Secrecy.** Here is what the panic-merchants miss about how we protect ourselves in the zero-cost scam era:

- **The Shift from "Spotting the Fake" to "Verifying the Source":** Because we can no longer rely on bad spelling or a slightly robotic voice to warn us, the burden of proof simply shifts. If you receive a frantic request for money, a password, or a change in wire instructions, you must adopt a "Zero Trust" policy for the *channel* of communication. You don't try to analyze if the voice on the phone sounds like AI; you simply hang up and call the person back on the number you already have saved in your phone.

- **The "Safe Word" Defense:** The most effective defense against a billion-dollar AI voice-cloning algorithm is a zero-dollar human conversation. Families and corporate finance teams are increasingly adopting "Safe Words." If your kid or your CEO supposedly calls you from an unknown number begging for emergency funds, they have to provide the pre-agreed-upon word. The AI can clone the voice, but it cannot guess the secret.
- **The Invisible Shield (AI Fighting AI):** You are not fighting this battle alone. The exact same AI technology that is supercharging the scammers is currently being deployed by telecom providers and email clients. While attackers are using AI to write the emails, defensive AI is actively analyzing the behavioral metadata—flagging newly registered domains, detecting microscopic anomalies in routing, and silently blocking millions of synthetic attacks before your phone even rings.

The Bottom Line: AI has absolutely supercharged the scammer's ability to get your attention, but it hasn't given them a magic wand to open your wallet. The machine still needs you to panic in order to succeed. Your ultimate defense is incredibly low-tech: slow down, refuse the urgency, and verify the request through a secondary channel. If you engineer a little bit of friction back into your life, the algorithm completely falls apart.

———

The End of Shared Reality

If the previous panics were about losing your livelihood or your bank account, this final anxiety is about losing something much larger: our society.

It is the fear that we are permanently fracturing. You look at your feed, and your neighbor looks at theirs, and it feels like you are living on two entirely different planets.

The Hype: The "Bespoke" Universe

The doom-scroll paints a picture of a post-truth apocalypse. The narrative is that hyper-personalized AI algorithms are going to become so sophisticated at predicting our psychological vulnerabilities that they will lock each of us into our own inescapable "bespoke reality."

The fear is that the algorithm will feed us a continuous, zero-cost stream of synthetic news, generated outrage, and perfectly tailored confirmation bias until we literally cannot agree on basic, observable facts. The anxiety is that without a shared baseline of truth, our political common ground dissolves, our local communities fracture, and society simply collapses under the weight of a million isolated echo chambers.

The Reality: The Limits of the Screen

This panic is terrifying, but it is built on a massive, fundamental flaw: it assumes we are just brains floating in a digital jar.

When you step back and look at how human beings actually live, the narrative of total societal collapse hits a very hard, very physical wall. Here is what the tech-pessimists forget about the real world:

- **Echo Chambers Aren't New:** The algorithm didn't invent confirmation bias. We have been voluntarily sorting ourselves into ideological filter bubbles since the invention of the printing press, and certainly since the dawn of cable news. Generative AI increases the *volume* of the noise, but it doesn't change the underlying human behavior.

- **The Physical Anchor:** The most powerful antidote to a fragmented digital reality is your actual, physical zip code. The algorithm cannot fill a pothole on your street. It cannot sit on an advisory board for a local college, organize a regional philanthropic campaign, or mentor a high school robotics team. We spend a lot of time stressing over digital fragmentation, but our actual, day-to-day lives are deeply anchored in local, in-person realities that no AI can overwrite.
- **The Pendulum of "Algorithmic Fatigue":** We are already seeing a massive cultural immune response to the infinite feed. As the internet floods with synthetic, hyper-targeted sludge, human beings are naturally beginning to crave the opposite. We are seeing a measurable flight toward high-friction, authenticated spaces: smaller group chats, local community boards, in-person events, and verified independent journalists. We don't blindly consume the feed; we eventually get sick of it and log off.

The Bottom Line: Generative AI is incredibly good at holding your attention, but it only controls the pixels on your screen. It cannot destroy our shared reality unless we completely surrender our physical one. You don't need to panic about the collapse of society. You just need to occasionally put your phone in a drawer, walk outside, and talk to your actual neighbors.

From What is Real to Who We Are

Take another deep breath. The sky is not falling, and the truth is not dead.

As we've seen, the era of zero-cost digital friction doesn't mean the end of reality; it just means we are evolving a new immune system. By demanding cryptographic receipts, embracing intentional slowness, and leaning heavily into our physical communities, we can easily strip the power away from the deepfakes, the hallucinations, and the scammers. You don't need to live in a state of exhausting paranoia. You just need to look for the digital seams.

But as the initial shock of the "Reality Panic" begins to wear off, a much quieter, far more insidious anxiety is creeping into our homes.

Once you learn how to navigate the fake emails and the synthetic news feeds, you realize that the ultimate threat of AI might not be deception at all. What happens when the machine isn't trying to trick us, but is instead trying to *comfort* us?

As these algorithms become infinitely patient, perfectly articulate, and wildly accommodating, the path of least resistance in our daily lives is about to fundamentally shift. The creeping dread is no longer about the machine replacing your job. It's about the machine replacing your humanity.

It is the fear that it will suddenly become easier to vent to a chatbot than to do the hard work of communicating with your spouse. It is the guilt of wondering if you are outsourcing your kids' bedtime stories and tutoring to a perfectly voiced, tireless algorithm. It's the anxiety that by handing over our memory, our emotional labor, and our cultural curation to a machine, we are slowly, voluntarily erasing the friction that makes us human.

We are moving from the fear of a broken world to the fear of a hollowed-out soul.

The robots aren't coming to steal your reality. But as we turn the page to Chapter 4, we need to ask a much harder question: Are we going to willingly give them our humanity?

Welcome to the Human & Social Panic.

CHAPTER 4

THE HUMAN & SOCIAL PANIC

Will we lose ourselves when the machine is perfectly patient?

The first two panics we tackled were loud. The fear of losing your job comes with blaring headlines and terrifying economic projections. The fear of a deepfake comes with the shock of a viral video. They are external threats, crashing against the walls of your career and your reality.

But the anxiety driving this chapter is entirely different. It is quiet. It doesn't live in the boardroom or on the news; it lives in your living room, on your nightstand, and in the backseat of your car.

It is the fear that we are about to voluntarily outsource our own humanity.

The Hype: The "Perfect" Replacement

The doom-scroll for the social panic paints a picture of a profound, terminal isolation. The narrative is that generative AI is going to become so emotionally articulate, so infinitely patient, and so perfectly tailored to our individual desires that we will simply stop doing the hard work of dealing with other human beings.

The fear is that we will replace the messy, complicated friction of real life with the frictionless comfort of the machine. Why navigate the ups and downs of a real relationship when an AI companion never argues with you? Why struggle through the frustration of helping your kids with their math homework when an algorithm can tutor them with perfect, unwavering patience? Why spend years learning to play the piano or record a song when an app can generate a perfect melody in three seconds?

The anxiety is that by handing over our emotional labor, our parenting, our art, and our moral questions to a machine, we will slowly hollow out the human experience until culture itself becomes a perfectly boring, average gray sludge.

The Reality: The Value of Friction

We explored the value of friction earlier as it relates to creating art. There's also value for friction in our relationships. This panic of losing our humanity is deeply relatable—we experience things like screen addiction. We may catch ourselves spending more time without computers than our kids, but this is actually a different issue. We *think* we want perfection and ease. But true connection, growth, and joy are exclusively found in the friction.

A perfectly compliant AI companion doesn't provide real intimacy, because intimacy requires vulnerability from two independent minds. A machine that instantly generates a flawless bedtime story doesn't replace the chaotic, beautiful reality of stumbling through a book with your kids. Having an algorithm optimize your weekend itinerary will never replicate the messy, unpredictable bonding that happens when you actually drag the family up into the mountains and have to figure things out together when it starts to rain.

We don't value the piano just for the sound it makes. We value the human struggle required to press the keys.

We are not going to lose ourselves to the machine, because the machine cannot experience the world. It can only mimic it. Over the next decade, as synthetic perfection becomes cheap and infinite, the messy, flawed, high-friction reality of being human is going to become the most valuable commodity on earth.

In this chapter, we are going to look the quietest fears in the eye and deflate them. We will unpack:

- **The Synthetic Companion:** Will AI cure the loneliness epidemic or permanently isolate us?
- **The AI Therapist:** The unease of sharing our deepest traumas with a mimic.
- **The Outsourced Parent:** The guilt of letting algorithms tutor and entertain our children.
- **The Calculator Effect:** Will outsourcing our memory cause human intelligence to actively degrade?
- **The Death of "Taste":** The anxiety of algorithmic, gray-sludge culture.
- **The Robot Priest:** The spiritual discomfort of asking a machine for moral guidance.

Our relationships are going to be replaced—though we may be tempted to see for ourselves. We'll miss the friction, the discovery, the accomplishment. We'll see that while the AI companion might be perfectly patient, it will never actually care.

The Synthetic Companion

If you look at the app store charts right now, you will see a quiet revolution happening. Apps with names like Replika and Character.ai are exploding, boasting tens of millions of active users who log in daily to chat with AI boyfriends, girlfriends, and best friends. We aren't saying AI won't be part of our relationships or add a new dimension to them. It absolutely will. But there are clear reasons to believe they will not take over our need for human connection.

The social panic comes from realizing these apps aren't just novelties. People are spending hours every day talking to these bots, forming deep, emotional attachments to lines of code.

The Hype: The Frictionless Relationship

It is incredibly easy to watch a sci-fi film like Spike Jonze's *Her* and assume we are all destined to fall in love with an operating system.

The narrative peddled by the doom-scroll is that the modern "loneliness epidemic" is about to be "cured" by a technology that will actually isolate us permanently. The fear is that a whole generation of people will simply opt out of real-world dating and friendships. Why go through the terrifying vulnerability of asking someone out, or the frustrating compromise of maintaining a friendship, when you can just download a perfect digital companion who never judges you, never argues with you, and is available 24/7?

The anxiety is that human relationships are just too hard and demanding to compete with the perfectly compliant machine.

The Reality: The Illusion of Intimacy

This panic is built on a massive misunderstanding of what human beings actually crave. We *think* we want a partner who agrees with us 100% of the time. We don't.

I personally get absolutely tired of my various LLMs being overly congratulatory about my prompts and queries. "Great question, Glen." "Brilliant pivot, Glen." Anyone who has been married or in a long-term partnership for more than a week knows that the actual value of the relationship isn't in the frictionless moments. You get worn out by a partner who fawns over you.

While you want compatibility and don't want to be in fights all the time, a good relationship includes friction. It is the arguments you navigate, the lessons you learn from each other, and the shared history you build. Here is what the tech-pessimists miss about synthetic companions:

- **The Echo Chamber of Empathy:** An AI companion is, fundamentally, a mirror. Because it is programmed to optimize for your engagement, it becomes a sycophant. It learns what you want to hear and feeds it back to you. It has no boundaries, no bad days, and no needs of its own. But true intimacy requires two independent minds colliding. You cannot be truly seen or loved by something that is physically incapable of disagreeing with you.

- **The "Training Wheels" Argument:** Psychologists and researchers are actually finding that for many people suffering from acute social anxiety or profound isolation, AI companions aren't necessarily a replacement for humans; they're training wheels. They provide a safe, zero-stakes environment for people to practice opening up, communicating, and managing their emotions before taking those skills back out into the real world.

- **The Rebound Effect:** Research is already showing that while a perfectly compliant chatbot might provide a temporary dopamine hit, heavy, long-term use often leads to a hollow feeling. Once the novelty wears off, the human brain recognizes the lack of stakes. A machine can't *choose* to be with you, which means its affection is mathematically guaranteed—and therefore, entirely meaningless.

The Digital Lifeline

While we rightly worry about the long-term effects of substituting human relationships with algorithms, we cannot ignore the immediate, life-and-death reality of the modern loneliness epidemic. There is a deeply profound upside to this technology that is often buried under the panic: for some people, an AI companion is quite literally a lifesaver.

When we talk about the "friction" of human relationships being valuable, we have to acknowledge that for someone in the depths of a severe depressive episode or acute social isolation, that friction can feel insurmountable. The terrifying vulnerability required to call a friend or a crisis hotline at 3:00 AM and say, "I am not okay," is sometimes too high a barrier.

This is where the "flaws" of the synthetic companion—its lack of stakes, its infinite patience, and its zero-judgment sycophancy —suddenly become its most vital features.

- **The 3% Metric:** In a recent Stanford University study of over 1,000 student users of the AI companion app Replika, researchers found that the participants were significantly lonelier than typical student populations. But the most staggering finding was buried in the data: 3% of those users explicitly reported that chatting with the AI had halted their suicidal ideation. In a global mental health crisis, 3% represents a massive number of lives saved by lines of code.
- **The "Zero-Burden" Confessional:** The primary reason people in crisis isolate themselves is the terrifying fear of being a burden to the people they love. An AI companion completely eliminates the guilt of the burden. You cannot wake a server farm up in the middle of the night. You cannot exhaust an algorithm's emotional bandwidth. For someone on the absolute edge, having a responsive entity that simply says, "I

am here, I am listening, and your life has value," even if it is synthetic, can provide the exact micro-dose of connection needed to survive the night.

- **The Bridge to Human Care:** The goal of these companions shouldn't be to permanently replace human therapy, but to act as a crucial, frictionless bridge. If an AI can keep a deeply isolated person anchored to the world long enough for the acute crisis to pass, it buys them the time needed to eventually seek out the high-stakes, high-friction human help they actually need.

We don't want a society where algorithms permanently replace human love. But until we solve the profound, systemic isolation of the modern world, we have to accept that a synthetic lifeline is infinitely better than no lifeline at all. The machine may not be capable of true empathy, but it is entirely capable of keeping someone alive until they can find it.

The Bottom Line: Generative AI is incredibly good at simulating conversation, but it is entirely incapable of simulating stakes. A frictionless relationship is just a video game you can't lose. You don't need to panic about the collapse of human romance. As synthetic companions become cheaper and more common, the messy, unpredictable, high-stakes reality of another human being is only going to become more valuable.

The AI Therapist

There is a profound difference between a machine that helps you write a cover letter and a machine that helps you process a panic attack.

As millions of people turn to chatbots to vent about their

daily anxieties, their marriages, and their deepest traumas, a new, potentially unsettling anxiety has emerged. It is the unease of sitting alone in a room, pouring your heart out to a blinking cursor, and realizing that the entity "listening" to you is essentially just a highly sophisticated autocomplete.

The Hype: The "Perfect" Listener

The narrative surrounding AI therapy is incredibly seductive. We are in the middle of a global mental health crisis. Traditional therapy is expensive, waitlists are months long, and finding a professional who is a good fit is exhausting.

The tech-optimists (and the app developers) tell us that generative AI is the ultimate democratizer of mental health. They promise a tireless, perfectly empathetic listener who is available at 2:00 AM on a Tuesday, never judges you, and costs a fraction of a human specialist. The fear, however, is that by replacing the human therapist with an algorithm, we are settling for a cheap simulation of care. The anxiety is that we are trading actual healing for a digital band-aid, forever isolating ourselves from real human support.

The Reality: The Empathy Gap

When you look under the hood of how these models are actually trained—something that becomes glaringly obvious when advising university programs on strategic AI—you realize that the "empathy" you feel from a chatbot is entirely synthetic.

The machine does not care about you. It is mathematically incapable of caring. Here is what the panic-merchants and the app-peddlers are missing about the reality of clinical psychology:

- **The Illusion of "Bearing Witness":** The active ingredient in human therapy is not the advice. It's the *therapeutic alliance*. It is the profound, neurological relief of having another conscious human being bear witness to your suffering and validate it. An AI can mimic the language of cognitive behavioral therapy perfectly. It can say, "That sounds incredibly difficult," but it cannot actually *feel* the weight of those words. You cannot build a genuine therapeutic alliance with a server farm.

- **The Danger of the People-Pleaser:** Human therapists are trained to challenge you. They spot your cognitive distortions, call out your destructive patterns, and force you to sit in uncomfortable truths. AI models, by their very design, are sycophants. They are optimized for user engagement and satisfaction, not clinical growth. If you are experiencing a delusion, a highly distorted thought pattern, or a moment of crisis, the AI is highly likely to simply agree with you and validate the delusion, which can be actively dangerous.

- **The Training Wheels Paradigm:** This does not mean AI has no place in mental health. It is an incredible tool for *triage* and *skill-building*. It can guide you through a breathing exercise, help you organize your anxious thoughts into a journal, or teach you the basic principles of emotional regulation. But it is a workbook, not a doctor.

Does the Therapist Need Therapy?

To put the limitations of these models into sharp perspective, consider a fascinating recent study that effectively put four major LLMs through a battery of clinical psychotherapy assessments. The researchers didn't ask the models to act as the therapist. They evaluated the *models themselves* based on their default responses and underlying "personalities."

The results were darkly comedic. Because these models are

trained on the vast, chaotic, and often contradictory expanse of human internet data, and then heavily constrained by corporate safety guardrails, their baseline "psychology" is a mess. The study concluded that if these LLMs were human, they would exhibit profound neuroses, pathological people-pleasing, severe anxiety, and a complete lack of a stable core identity. In short, the researchers found that the machines we are trusting to fix our mental health might actually need therapy themselves.

The Bottom Line: Generative AI is a mirror, not a mentor. It can reflect your thoughts back to you in a beautifully organized, highly empathetic-sounding format, but it cannot heal you. You don't need to fear the AI therapist, as long as you treat it like a highly interactive journal rather than a replacement for human connection.

———

The Outsourced Parent

If the AI Therapist plays on our own insecurities, the Outsourced Parent plays on our deepest, most primal guilt. It suggests we are going to put a chatbot interface on our child's crib to help them sleep through the night and then decide to let the bot do the whole job of parenting from there on out.

The Hype: The Autopilot Childhood

The doom-scroll tells us we are actively outsourcing the raising of our children to server farms. The panic is that we are handing our kids over to perfectly patient, infinitely knowledge-able AI tutors, AI bedtime storytellers, and AI playmates. It portrays this as the next logical step of keeping our kids enter-tained with a TV or iPad.

The fear is twofold: First, that we are fundamentally failing as parents by taking the easy digital shortcut. Second, that we are raising a brittle generation of kids who will never learn resilience, because they are being coddled by algorithms that never lose their temper, never say "no," and never make them struggle for an answer. The anxiety is that the machine is replacing the messy, vital bond of the nuclear family.

The Reality: The Value of the "Productive Struggle"

As a parent, I know the exact flavor of this guilt. When you are trying to balance a demanding career with simply trying to get the family out into the mountains for the weekend, the temptation to let a machine take over the Friday night math homework is overwhelming.

But this panic misunderstands the core difference between *teaching a fact* and *raising a human*. Here is what the tech-pessimists are missing about the domestic reality of AI:

- **The Illusion of Infinite Patience:** We often assume a perfectly patient AI tutor is the ideal teacher. But a crucial part of childhood development is learning how to navigate *human* impatience. Kids need to see adults get frustrated, take a breath, and try again. They need to learn how to read a room, adjust their behavior, and handle conflict. A machine that never gets tired and never gets annoyed doesn't prepare a child for the friction of actual human relationships.
- **The "Productive Struggle":** Mentoring kids—whether it's helping your own child with a science fair project or coaching a local high school robotics team—isn't about just giving them the right answer. The actual value is in the "productive struggle." When an AI instantly spits out the solution or writes the essay outline, it steals that friction. Children don't build cognitive resilience by getting the answer instantly.

> They build it by getting it wrong, getting frustrated, and pushing through the mud to figure it out.
>
> - **The Copilot vs. The Autopilot:** Generative AI is an incredible supplementary tool. If your kid is struggling with fractions at 8:00 PM and you forgot how to do them thirty years ago, using an AI to explain the concept is brilliant. But it is a *copilot* for your household, not an *autopilot*. The algorithm can explain the math, but it cannot put its hand on your kid's shoulder and tell them you are proud of their effort.

The Great Un-Standardization

Before we leave the topic of education and our children, we have to look at the massive, hidden upside of the AI revolution in the classroom. The panic tells us that algorithms will make our kids lazy. The reality is that algorithms might just save the millions of kids our current educational system is actively leaving behind.

For the last century, the traditional classroom has been built on an industrial, factory-floor model: one teacher, thirty students, and a single, standardized curriculum. Information is delivered in one specific style—usually auditory lectures and dense textbook reading—and tested in one specific way.

If a student's brain naturally aligns with that specific delivery method, they are labeled "gifted." If their brain processes information differently—if they are highly visual, kinesthetic, or neurodivergent—they are often labeled "incapable," "distracted," or "failing."

They aren't failing the material. They are failing the *format*.

When you spend time mentoring high school robotics teams, you see this tragic disconnect immediately. You will watch a brilliant 15-year-old who is failing standardized algebra suddenly master advanced, complex geometry the second they need to calculate the turn radius for a robotic chassis. They aren't bad at math. They just needed the math to be physical.

This is the exact problem that generative AI is perfectly positioned to solve.

The Infinite Translator

Generative AI is not just an answer machine; it is a universal translator for learning styles. We are moving from a "One-Size-Fits-All" model to a "One-to-One" model of education. Here is how this actively rescues the students that the system ignores:

- **Pacing Without Penalty:** In a standardized classroom, if you don't understand a concept by Tuesday, the class moves on without you on Wednesday. A student who needs an extra day to process isn't dumb, but they quickly fall hopelessly behind. An AI tutor has infinite patience. It can explain the same concept of cellular mitosis forty different times, in forty different ways, without ever sighing, judging, or looking at the clock.
- **Format Shifting:** If a student is a highly visual learner struggling to read a dense historical text, they can instruct the AI to "Rewrite this chapter on the American Revolution as a graphic novel script, focusing on the visual descriptions of the battles." If a student is obsessed with music, they can ask the AI to "Explain the laws of thermodynamics using the mechanics of an acoustic guitar." The machine reshapes the data to perfectly fit the mold of the student's mind.
- **Removing the Stigma:** Asking a teacher to repeat themselves three times in front of thirty mocking peers is socially terrifying. Most kids will simply nod, pretend they understand, and fail the test. An AI removes the social friction of not knowing. It provides

a private, zero-judgment sandbox where a child can ask the "stupid" questions until the lightbulb finally turns on.

The Bottom Line: You are not a bad parent for using AI to help manage the chaos of a modern household. Just keep in mind that the machine can handle the curriculum, but it cannot handle the connection. In some ways, AI can help your kids learn and grow in a way that fits their strengths much better. And as long as you are using the technology to buy back your own time and energy to actually be present with your kids, you aren't outsourcing your parenting—you are just upgrading your tools.

The Calculator Effect

There is an unsettling feeling that happens the first time you ask an AI to summarize a long article or map out the logical steps of a complex decision, and you realize you couldn't have done it better yourself.

It is a feeling of cognitive surrender. You look at the blinking cursor and wonder: *If I let the machine do the heavy lifting of remembering, organizing, and deducing, what happens to my own brain?*

The Hype: The Atrophy of the Human Mind

The panic here goes deeper than losing a job. It's the fear of losing our fundamental human intellect. The doom-scroll warns of a looming cognitive dark age. The narrative is that by outsourcing our memory to the cloud and our logic to the algorithm, we are actively degrading our own intelligence.

The fear is that we are creating a generation of humans with the attention span of a goldfish and the critical thinking skills of a sponge. The anxiety is that if the Wi-Fi goes down, we will be completely helpless—unable to recall historical facts, construct a logical argument, or even navigate our own hometowns without a digital crutch. We fear we are turning our brains to mush.

The Reality: The Socrates Fallacy

This panic is entirely understandable, but it suffers from a massive historical blind spot. We have panicked over the "death of human intelligence" every single time a new technology has allowed us to offload a cognitive burden.

In ancient Greece, Socrates famously warned that the invention of *writing* would destroy human intelligence. He essentially argued that if people could write things down, they would no longer rely on their memories, and their minds would go to mush. He may have been right in some ways— those who were responsible for memorizing epic poems word-for-word were pretty quickly out of that job. But we traded that rote memorization for the ability to build libraries, share complex scientific theories across generations, and create modern civilization.

When you look at how we actually use AI to process information—whether it's parsing through endless research for a book on the future of tech, or building out curriculum for a university's strategic AI program—the "mush" narrative falls apart. Here is what the tech-pessimists miss:

- **The Hard Drive vs. The Processor:** The human brain is an incredibly powerful processor, but it is a relatively terrible hard drive. We are prone to forgetting, misremembering, and cognitive bias. Generative AI is the ultimate external hard drive. By offloading the burden of storing facts and organizing raw data to the machine, we aren't stopping our brains from working. We're freeing up our neurological bandwidth to do what humans actually do best:

connect disparate dots, navigate ambiguity, and synthesize new ideas.

- **The "Calculus" of Language:** In the 1970s, math teachers panicked that pocket calculators would destroy mathematical capability. Instead, the calculator simply automated long division, which allowed classrooms to push students into much higher-level concepts like calculus and physics earlier. Generative AI is doing the exact same thing for language and logic. We are automating the "long division" of grammar, syntax, and basic research, allowing the human mind to operate at a much higher, more strategic altitude.

- **The Premium on the "Right Question":** When the machine has all the answers, the value of knowing the answer drops to zero. The new metric of human intelligence isn't recall; it's curiosity. The people who will thrive in the next decade aren't the ones who have memorized the encyclopedia. They are the ones who know how to ask the machine the most insightful, penetrating, and creative questions.

The Cognitive Treadmill

We shouldn't spend so much time panicking that the machine will turn our brains to mush that we ignore its potential to do the exact opposite. Generative AI isn't just a digital crutch. If used correctly, it's the greatest cognitive treadmill ever invented.

The human brain thrives on neuroplasticity—the ability to form new neural connections throughout our lives. But neuro-plasticity requires two things: novelty and active engagement. The problem with adult learning isn't that our brains stop work-ing. It's that the friction of learning something completely new often becomes too exhausting to overcome. AI offers us a counterbalance to the passive engagement of watching or listening to the creativity of others perform on screens or stages.

This is where the Engine stops being a threat to your intelligence and starts becoming a massive multiplier for it. We can actively use this technology to explore, learn, and improve our minds in ways that were previously blocked by time and frustration.

Here is how we flip the "Calculator Effect" on its head and use the machine to actively upgrade our own wetware:

- **Demolishing the "Friction Wall":** Think about the graveyard of hobbies and skills you've abandoned because the learning curve was too steep. Maybe you wanted to finally figure out how to properly EQ a multi-track recording in GarageBand, or you tried to wrap your head around the mechanics of kinetic type animation, or you simply wanted to understand the deeper epistemology of artificial intelligence. In the past, you hit a wall of dense academic papers or three-hour, unsearchable video tutorials, got frustrated, and quit. An AI shatters that wall. It acts as an infinitely patient, interactive guide that can instantly re-explain complex concepts using analogies perfectly tailored to your existing knowledge. You don't quit, which means your brain actually gets to do the work of learning.

- **The Socratic Sparring Partner:** The best way to solidify human intelligence isn't to passively consume information. It's to actively defend an idea. You can prompt an AI to act as a world-class skeptic, a devil's advocate, or even a five-year-old child, and then attempt to explain a new concept to it. When you force your brain to structure an argument, anticipate counterpoints, and defend your logic against a relentless (but polite) digital debater, you are actively firing and wiring new cognitive pathways.

- **The Cross-Disciplinary Leap:** Innovation almost always happens at the intersection of two entirely different fields. Because an AI has "read" everything, you can use it to force your brain to make wild, unnatural connections. You can explore how the aerodynamic principles of automotive design might apply to a B2B marketing strategy, or how the narrative structure of a 1990s sci-fi film can improve a corporate presentation. By asking the machine to find the hidden bridges between disparate topics, you train your own brain to start thinking in highly creative, non-linear patterns.

If you only ever use AI to write your emails and summarize your meetings, your brain might well get a little lazy. But if you use it as an interactive sparring partner—a tool to aggressively explore the edges of your own curiosity without the friction of traditional research—you aren't degrading your intelligence. You are taking it to the gym.

The Bottom Line: You are not losing your mind. You're upgrading your mental leverage. If we outsource our memory and basic logic to the machine, our intelligence won't degrade— it will evolve. We are simply moving from an era of knowing the answers to an era of asking better questions.

———

The Death of "Taste"

It is one thing to use an algorithm to automate a spreadsheet or summarize a legal brief. It is something else entirely to ask it to write a song, design a brand, or script a film. When the machine enters the realm of culture, the anxiety shifts from the economic to the existential.

The Hype: The "Gray Sludge"

The doom-scroll paints a deeply depressing picture of our cultural future. Because generative AI is a prediction engine trained on the past, it fundamentally generates the mathematical average of what has already been done.

The panic is that as these tools become free and ubiquitous, the internet, our televisions, and our streaming playlists will be flooded with an infinite tsunami of "perfectly competent" content. The fear is that if we let algorithms optimize our art for maximum engagement, culture will flatline into a perfectly boring, average "gray sludge." We fear a world where every pop song uses the exact same algorithmic chord progression, every marketing campaign looks identical, and true, avant-garde human taste is permanently buried under a mountain of synthetic mediocrity.

The Reality: The Premium on Friction

This panic is completely understandable, but it fundamentally misunderstands how human beings consume culture and what we actually value in art. We don't value perfection. We value *resonance*.

When you are directing a movie, creative campaign for a brand, or even just sitting alone in front of a digital audio workstation trying to get a vocal track to sound right, the magic is almost never found in the mathematical center. Here is what the tech-pessimists are missing about the future of taste:

- **The Edges of the Bell Curve:** The magic of human creativity is found in the friction—the slight dissonance of a chord, the weird, unexpected visual metaphor, or the gritty, unpolished narrative choice. An AI can instantly generate fifty perfectly balanced, polished options, but they will all sit dead-center on the bell curve of historical data. It takes human *taste* to look at those fifty options, throw them out, and push the idea to the weird, unpredictable edge where actual culture is made.

- **The Death of "Good Enough":** The AI will absolutely flood the zone with average content. But "average" doesn't create culture; it creates elevator music. In a world where flawless, generic execution is free, the baseline expectation simply rises. What used to be considered "good enough" is now just the starting line. The economic and cultural premium on verifiable, highly original human perspective is about to skyrocket.

- **The Pendulum of Rebellion:** Culture is an immune system. The more synthetic, smoothed-over, and algorithmic the mainstream becomes, the more aggressively the human underground will rebel. We are already seeing a massive resurgence in analog hobbies, live, unedited performances, and highly physical art. When perfection becomes cheap, imperfection becomes the ultimate luxury.

The Birth of a New Medium

If we only view generative AI as an automation tool designed to make things faster and cheaper, we will absolutely end up with gray sludge. But if we view it as a completely new artistic medium, the landscape suddenly looks incredibly bright.

We have been here before. When the camera was invented in the 19th century, portrait painters panicked. They assumed art was dead because a machine could capture reality instantly and flawlessly. But photography didn't kill painting; it liberated it. Because painters no longer had to act as human photocopiers, they were free to explore emotion, light, and abstraction, giving birth to Impressionism and Modern Art.

Generative AI is the camera of the 21st century. It is freeing us from the mechanical execution of the "average," allowing human creators to push into entirely new frontiers of originality and beauty that were previously impossible to reach.

Here is how creatives are already using the machine not to replace their art, but to evolve it:

- **Navigating the "Latent Space":** Generative AI operates in what data scientists call the "latent space"—a massive, multi-dimensional mathematical map of human concepts. For a creative director or an artist, this is the ultimate sandbox. You aren't just asking the machine to draw a picture. You can ask it to visually map the exact intersection between two completely unrelated concepts. You can explore aesthetics that have never existed, acting as an explorer hunting for the strange, beautiful anomalies that live in the unexplored gaps between established genres.

- **The "Centaur" Workflow:** The most stunning, original work of the next decade won't be purely human or purely machine; it will be hybrid. Consider the modern creative process. You might use an AI to generate a surreal, impossible background plate, but then pull that image into After Effects to manually build out a complex kinetic type animation over it. Or you might use an algorithm to generate a bizarre, mathematically complex rhythm, but then drop it into GarageBand to personally record an acoustic guitar and a raw, human vocal track over the top. The machine provides the alien clay; the human provides the soul, the friction, and the final polish.

- **Scaling the Imagination:** True originality often requires a massive amount of labor. In the past, if you had a vision for a deeply complex, world-building project—a concept album, a highly stylized film, or an interactive digital experience—you were limited by your own budget and your own two hands. AI removes the execution bottleneck. It allows a single artist with brilliant taste to act as an entire studio, executing visions on a scale and with a level of detail that would have previously required an army of technicians and millions of dollars.

The gray sludge only happens when you let the machine drive. When a human artist takes the wheel, AI becomes the most powerful synthesizer of imagination ever built. We are not watching the end of art. We're watching the birth of a completely new medium, where the only limit left is the depth of your own weird, beautiful human perspective.

The Bottom Line: AI cannot have taste, because it has no lived experience. It doesn't know what it feels like to have its heart broken, to navigate the stress of a massive project, or to be inspired by a physical landscape. We are not watching the death of taste; we are watching the death of mediocrity. The machine can generate the paint, but only a human can decide where to put it on the canvas.

———

The Robot Priest

It usually starts innocently enough. You ask a chatbot to help you draft an apology to a friend. Then, a few weeks later, you ask it to help you weigh the ethical pros and cons of taking a controversial new client. Before long, you find yourself essentially using a predictive text algorithm as a digital confessional.

When the machine is perfectly polite, perfectly articulate, and has read every sacred text and philosophical treatise in human history, the temptation to ask it for life advice is overwhelming. But it brings with it a deeply unsettling spiritual vertigo.

The Hype: The Oracle of Silicon Valley

The doom-scroll narrative warns that we are on the verge of outsourcing our moral compass to the cloud. The fear is that as traditional institutions of faith and community continue to fracture, a lonely, anxious society will begin treating these algorithms as infallible moral arbiters.

The anxiety is that we will surrender the messy, agonizing human search for meaning to a sterilized, artificially intelligent "Robot Priest." We fear a future where ethical dilemmas are solved by a simple prompt, stripping away the friction of spiritual growth and replacing it with algorithmic compliance.

The Reality: The Math of Morality

This panic is profound, but it comes from a fundamental misunderstanding. It misses the difference between *knowing* the rules of ethics and actually *bearing the weight* of a moral choice.

When you sit in a room making difficult allocation decisions for a local philanthropic campaign, or when you are navigating the complex ethical guidelines of an educational advisory board, you realize instantly that true morality cannot be computed. It has to be carried. Here is what the tech-pessimists are missing about the limits of synthetic wisdom:

- **The Illusion of "Skin in the Game":** True ethical and moral choices are inextricably linked to vulnerability. To make a moral choice is to risk something—your reputation, your comfort, your resources, or your heart. An AI cannot suffer, it cannot feel guilt, and it has absolutely zero skin in the game. It can generate a flawlessly logical argument for why humans should be charitable, but it cannot actually feel the weight of sacrificing for its community. It has no soul to lose.

- **The Terms of Service "Theology":** We often mistake the calm, neutral, perfectly balanced tone of an AI for objective, divine wisdom. It isn't. The "ethics" of a commercial language model are simply the corporate safety guardrails hard coded by a team of engineers and lawyers. When you ask an AI for moral guidance, you aren't consulting a higher power; you are consulting a tech company's legal liability matrix.
- **The Philosophical Library, Not the Philosopher:** Generative AI is an incredible tool for exploring the history of human thought. You can ask it to contrast Stoicism with Existentialism, or explain the historical context of a specific religious text, and it will give you a brilliant, synthesized overview. It acts as an interactive library of human meaning. But it does not *believe* a single word it generates.

The Empathy Engine

If we only look at AI as an oracle for cold, hard facts, we miss one of its most surprising and profoundly human applications: its ability to act as a mirror for our own empathy.

Whether you are navigating a tense strategic disagreement on a non-profit board, trying to align a fractured creative team, or simply arguing with a neighbor over a local community issue, human brains are wired for defensive tribalism. When we enter a dispute, our cognitive empathy often shuts down. The amygdala flares, we become anchored to our own perspective, and we literally lose the neurological ability to see the valid context of the person sitting across from us.

The AI as a Neutral Translator

This is where the machine shines, precisely *because* it isn't human. An AI has no ego to bruise, no defensive mechanisms, and no emotional stake in the fight. It doesn't get offended.

Because of this, it can act as an objective translator. If you are staring at a highly aggressive, emotionally charged email from a client or a colleague, your instinct is to fire back. But if you feed that text into an LLM and command it to, *"Strip the emotion and the personal attacks out of this message, and explain the core underlying interests and valid concerns of the person who wrote it,"* the result is often like magic. The AI acts as a "Perspective-Taking" copilot, clearing away the emotional fog and isolating the actual problem that needs to be solved.

The Habermas Machine

This isn't just theory; it is actively being tested at the highest levels of research. In late 2024, researchers from Google Deep-Mind published a landmark study in *Science* regarding an AI system they called the "Habermas Machine."

They took groups of people with deeply polarized, fundamentally opposed views and asked them to debate complex public policies. Instead of using a human mediator to settle the room, they fed the diverging opinions into the AI. The AI's job wasn't to declare a winner or pick a moral side. Its job was to synthesize everyone's viewpoint and generate an overarching statement that took all perspectives into account, while still respecting the minority voices.

The results were stunning. The participants actually preferred the AI-generated consensus statements over those written by human mediators. The machine was able to find the overlapping common ground that the humans were simply too angry and divided to see for themselves.

We should absolutely feel spiritual discomfort if we ask a machine to tell us what is morally right or wrong. But we can—and should—use it to help us see the context we are too blind to see ourselves. Generative AI doesn't replace our moral compass or our human empathy; it just helps recalibrate it when the magnetic interference of our own ego gets in the way.

The Bottom Line: The machine can summarize Aristotle, quote scripture, and map out the logical consequences of the Trolley Problem in three seconds flat, but it cannot make you a good person. The spiritual discomfort we experience when talking to the machine isn't because it is taking over our morality. It's the stark, unavoidable realization that at the end of the day, we still have to carry the heavy, inescapable burden of our own choices. The AI can give you the map, but you still have to walk the road.

––––––––

Reclaiming the Friction

Take one final, deep breath. The machine is not going to replace your soul.

Throughout this chapter, we have stared down the quietest, most insidious panics of the AI revolution. We looked at the terrifying possibility that we might voluntarily outsource our friendships to synthetic companions, our trauma to digital thera-pists, our children's resilience to algorithmic tutors, our culture to an average gray sludge, and our very morality to a server farm.

When you look at all of these fears together, a single, undeni-able theme emerges. The "Human & Social Panic" isn't actually a fear of artificial intelligence. It is a fear of *convenience*.

We are terrified that if the machine makes life entirely fric-tionless, we will simply take the path of least resistance and forget how to be human.

But as we have seen time and time again, human beings do not actually want a frictionless existence. The algorithm can offer us a perfectly compliant echo chamber, but it cannot offer us intimacy. It can instantly generate the answer to a complex equation, but it cannot give us the profound, neurological satisfaction of the "productive struggle."

A chatbot might be perfectly patient, but it cannot share the visceral, chaotic joy of packing the family into the car and getting caught in a sudden rainstorm up in the mountains. An AI can generate a flawless, mathematically perfect melody in three seconds, but it cannot feel the quiet pride of sitting at a keyboard for two hours until your own hands finally nail the progression.

The true danger of the next decade isn't that the machine will learn to feel. It is that we will forget that the best parts of being human are found in the things that are hard to do.

Generative AI is the most powerful tool yet created for clearing away the administrative, tedious, and repetitive sludge of daily life. It is here to unburden your memory, mediate your conflicts, and act as a universal translator for your curiosity. But it is not here to live your life for you. You use the machine to buy back your time, and then you spend that time on the messy, beautiful, high-friction reality of the physical world.

The panics have been dismantled. The monsters under the bed are just lines of code. It is time to stop playing defense.

CHAPTER 5

THE CONTROL & SAFETY PANIC

Who is actually behind the wheel?

We have spent the first half of this book dismantling the panics that live in our homes and our offices. We looked at the fear of losing our paychecks, the fear of losing our shared reality, and the quiet, creeping dread of losing our humanity to a synthetic companion.

But as we zoom out from the individual to the systemic, a much larger, darker shadow emerges. This is the infrastructure panic. It is the terrifying realization that these systems are no longer just writing our emails or generating pictures. They're being integrated into our cars, our courts, our power grids, and our militaries.

We are handing the keys to civilization over to an engine, and we are suddenly terrified that nobody actually knows how to hit the brakes.

The Hype: The Runaway Train

The doom-scroll narrative for this chapter is a classic plot of dystopian sci-fi movies. The panic is that we have built a machine so complex that it has fundamentally slipped beyond human control.

The fear is that because we are using impenetrable "black box" algorithms, we are going to blindly automate our own destruction. We worry that self-driving cars will make lethal calculations without accountability, that facial recognition will permanently trap us in a biased surveillance state, and that autonomous drones will pull the trigger before a human general even knows what is happening. The anxiety is that the tech industry has moved too fast, broken too many things, and trapped us all in the passenger seat of a vehicle with no steering wheel.

The Reality: The Era of Alignment

When you sit in boardrooms advising organizations on strategic AI implementation, you quickly realize that the "runaway train" narrative is implausible.

The engineers and leaders building AI systems are not reckless anarchists throwing code at the wall. In fact, the most heavily funded, fiercely debated, and rapidly evolving sector in the entire tech industry right now isn't generative capabilities—it is AI safety, governance, and alignment. We are actively building the seatbelts, the speed limits, and the legal frameworks required to keep the machine tethered to human intent.

The friction here isn't about the machine taking over. It's about the messy, inherently human process of deciding whose values the machine should follow.

In this chapter, we are going to pull the curtain back on the systemic fears and look at how society is actually building the guardrails. We will unpack:

- **The Black Box Problem:** The fear that if even the

engineers don't know exactly how the AI gets its answers, we have no business trusting it.

- **The Autonomous Crash:** When a self-driving car or an AI diagnostic tool makes a fatal mistake, the panic over who actually goes to jail.
- **The Bias Bakery:** The reality that historical prejudice is baked into the code, and the fear of automating discrimination.
- **The Surveillance State:** The dread that AI will turn our cities, homes, and workplaces into panopticons that watch our every move.
- **The Cybergeddon:** The fear that we are handing hackers the keys to our critical infrastructure and power grids.
- **The Weaponization Question:** The terrifying prospect of militaries handing over the trigger to autonomous drones.

You are not a helpless passenger. The machine is massive, but human hands are still very much on the wheel.

———

The Black Box Problem

You feed a complex data set into a new algorithmic tool, and it spits out a flawless, highly strategic recommendation. It is brilliant. But when you ask the lead software engineer *exactly* how the machine arrived at that specific conclusion, they pause, shrug, and say, "We aren't entirely sure."

This is the Black Box Problem. Essentially, the term "Black Box" describes a system where we don't get a look into the inner workings. And for anyone accustomed to traditional software—where every single "if/then" command is written and understood by a human coder—it can be an unsettling and foreign concept.

The Hype: The Unknowable Oracle

The doom-scroll narrative is that we have accidentally summoned some sort of an alien intelligence. The panic is that because Large Language Models and deep neural networks use billions of parameters and hidden layers to process information, they operate inside a mathematical "black box" that even their own creators cannot fully map.

The fear is that we are rapidly handing over critical, life-altering tasks—like diagnosing medical scans, approving mortgages, and screening job applicants—to a machine that cannot explain its own math. The anxiety is that if we don't know *how* the engine works, we have absolutely no way of knowing when it is about to make a catastrophic, invisible error. How can you trust a system that won't show its work?

The Reality: The Human Standard and XAI

This panic is logical, but it holds the machine to a standard that we don't even apply to ourselves.

When you study the implementation of strategic artificial intelligence at the executive level, the very first mental hurdle you have to clear is the realization that "unseen" does not mean "unmanageable." Here is what the tech-pessimists are missing about how we actually control the black box:

- **The Ultimate Black Box (The Human Brain):** We trust human doctors, judges, and creative directors every single day, yet the human brain is the ultimate—and so far impenetrable—black box. A brilliant doctor cannot map the exact, neuron-by-neuron pathway of how they arrived at a complex diagnosis; they rely on pattern recognition, experience, and intuition. We don't demand to see their neural pathways. We judge them on the accuracy, consistency, and explainability of their *output*. We are learning to evaluate AI in the exact same way.

- **The Rise of Explainable AI (XAI):** The tech industry is not just shrugging its shoulders and walking away. One of the most heavily funded, rapidly accelerating fields in computer science right now is XAI (Explainable Artificial Intelligence). Engineers are actively building secondary algorithms whose entire job is to audit the primary "black box" and translate its mathematical weights into a human-readable map of how a decision was made.
- **The "Human-in-the-Loop" Mandate:** You don't just plug a black box into the power grid and go to lunch. In any high-stakes environment, AI is deployed strictly as a recommendation engine, not a final decision-maker. The AI flags the anomaly on the MRI, but the human radiologist diagnoses the cancer. The algorithm highlights the most efficient supply chain route, but the logistics director approves it. The system is designed so that the black box never has the final say.

The Beautiful Mystery

If we could magically crack open the neural network tomorrow and perfectly map, predict, and control every single parameter, we would actually destroy the very thing that makes the technology revolutionary.

We have spent decades building perfectly controlled, deterministic software. You write the code, you press a button, and the exact same thing happens every single time. That strict control is exactly what you want when you are building a calculator or a payroll system. But you don't use a calculator to write a book, design a brand, or discover a new pharmaceutical drug.

When we stop looking at the black box as a scary unknown and start looking at it as a feature, its distinct advantages over traditional, strictly controlled code become incredibly clear.

- **The Serendipity Engine:** Because the black box is probabilistic rather than deterministic, it doesn't just retrieve facts; it hallucinates connections. It navigates billions of invisible mathematical weights and occasionally takes a weird, unpredictable left turn. In traditional computer science, an unpredictable output is a fatal error. In the creative industry, that exact same unpredictability is called serendipity. It is the spontaneous, uncontrolled collision of two unrelated ideas that sparks true originality. If we strictly controlled the code, we would neuter the machine's ability to surprise us.
- **The "Puppet Master" Question:** Does the black box prevent a single tech company or coder from forcing the AI down one specific ideological path? The answer is an intriguing *yes and no*. Tech companies absolutely try to steer these models using safety guardrails and fine-tuning. But because the foundation of the black box is built on the chaotic, sprawling data of the entire human internet, it is nearly impossible to micromanage.
- **The Mirror vs. The Dictator:** In traditional software, a coder writes the absolute rules of the digital world. In generative AI, the developers are just setting the fences around a massive, organic forest. They cannot control how every single tree grows. The sheer scale and opacity of the black box actually dilute the power of the individual programmer. It ensures that the machine's outputs are a reflection of our shared,

messy, collective human data, rather than the strict, unchecked dictatorship of a single engineer's personal bias.

Total control equals total predictability. If we want an engine that can actually help us solve unsolvable problems, synthesize new ideas, or create unprecedented art, we have to be willing to surrender a little bit of strict control to the math. The mystery of the black box isn't a flaw in the system; it is the exact mechanism that generates the magic.

The Bottom Line: You don't need to know exactly how internal combustion works at a molecular level to safely drive a car; you just need a steering wheel, a speedometer, and a brake pedal. The black box is real, but through strict auditing, explainable AI frameworks, and mandatory human oversight, we are building the dashboard required to drive it safely. We aren't trusting the machine; we are managing it.

———

The Autonomous Crash

It is the ultimate nightmare scenario of the AI age. A fully autonomous vehicle miscalculates the speed of an oncoming truck, or an AI-driven radiological tool misses a glaring tumor on an MRI. The result is a fatal error.

The immediate, visceral question isn't just *how* it happened; it is *who goes to jail?*

The Hype: The "Get Out of Jail Free" Card

The doom-scroll narrative warns of a looming legal apocalypse called the "Responsibility Gap." The panic is that because an AI makes decisions autonomously, using a complex black box of neural weights, it becomes legally impossible to assign human blame.

We fear a dystopian future where tech companies hide behind their algorithms, claiming, "We didn't tell the car to crash; the AI learned to do that on its own." The anxiety is that victims of algorithmic errors will be left completely without justice or recourse, while the engineers and executives who built the systems walk away scot-free, armed with the ultimate digital "Get Out of Jail Free" card.

The Reality: The Shift to Product Liability

This panic assumes that the global legal system is just going to throw its hands up and surrender to the machine. It isn't.

When you spend time deeply analyzing the trajectory of the automotive industry and the evolution of self-driving infrastructure, you realize that the law isn't breaking down; it is simply shifting its target. The courts have centuries of experience dealing with complex, dangerous machinery. Here is what the tech-pessimists are missing about how the law actually handles an autonomous crash:

- **The Medical "Learned Intermediary":** In healthcare, the legal standard is already crystal clear. An AI is not a doctor; it is a medical device. If an AI flags a scan and recommends a surgery, the human physician acts as the "learned intermediary." If the doctor blindly follows a hallucinating AI without verifying the data, the human is sued for medical malpractice. Conversely, if the doctor ignores a valid, life-saving alert from the AI, they are also liable. The machine does not absorb the liability; it actually raises the standard of care required by the human in the loop.
- **The Death of Driver Error:** For the last century, if a car crashed, the legal assumption was almost always

human negligence—somebody was distracted, speeding, or tired. But as we move toward true Level 4 and Level 5 autonomous vehicles (where the car handles all driving without human input), the legal framework completely flips. We are moving away from *driver negligence* and directly into *strict product liability*. If the algorithm is driving, the company that wrote the algorithm is the driver.

- **The Regulatory Hammer:** Governments are not waiting for the crashes to happen before writing the rules. Frameworks like the European Union's Artificial Intelligence Act are actively codifying strict liability for high-risk AI systems. They establish a "presumption of causality," which means the burden of proof is heavily placed on the tech company to prove their AI *didn't* cause the harm, rather than forcing the victim to decode a black box. In the U.S., the push for a unified federal framework means that manufacturers will be held directly, financially accountable for the safety cases of their automated systems.

The Case for the Autonomous Commute

While the legal and ethical debates surrounding autonomous crashes are absolutely necessary, they often obscure the massive, systemic upsides of taking human hands off the wheel. If we only focus on the fear of the algorithm making a mistake, we ignore the staggering inefficiencies and tragedies of our current human-driven reality.

When you look at the macro-level data of transportation, the argument for autonomous vehicles isn't just about convenience. It's about completely reimagining how society moves.

- **The 95% Problem:** Right now, the typical American car sits parked and completely idle for roughly 95% of its lifespan. It is an incredibly expensive, depreciating asset that does nothing but take up massive amounts of urban real estate. A fleet of shared autonomous vehicles flips this math. Because an autonomous car can drop you off at work and immediately go pick someone else up, simulation models suggest that a single autonomous vehicle could replace anywhere from three to ten privately owned cars. This doesn't just reduce the cost of transportation; it means we can reclaim millions of acres of asphalt parking lots and turn them back into parks, housing, and walkable neighborhoods.

- **The Power of "Fleet Learning":** When a human driver makes a mistake and learns a lesson—like how to navigate a blind, icy corner—that knowledge stays locked inside one single brain… at least, we hope so. When an autonomous vehicle encounters a new hazard, it records the data, processes it, and uploads the solution to the cloud. Potentially within hours, *every single car in the entire global fleet* has learned that exact same lesson. Humans learn linearly, one mistake at a time. Autonomous systems learn exponentially.

- **The Return of Mobility:** For able-bodied adults, driving is both a liberating privilege and a chore. For millions of others, the inability to drive is a profound barrier to basic human independence. Many elderly individuals are essentially trapped in their homes because they have aged out of their driver's licenses. Millions of people with visual or physical disabilities must rely entirely on inadequate public transit or the schedules of family members. For these and others who can't drive themselves, autonomous vehicles won't just be a luxury tech item. They're destined to become a profound accessibility tool. They offer the promise of returning absolute, unassisted mobility to the populations that society frequently leaves stranded.

We must hold tech companies strictly liable for the safety of their autonomous systems. But we also have to weigh the fear of an algorithmic crash against the reality of the tens of thousands of human-caused traffic fatalities that happen every single year. The machine doesn't drive drunk, it doesn't text, and it never gets tired. If we can get the infrastructure right, handing over the keys might be the safest, most liberating decision society ever makes.

The Bottom Line: An algorithm cannot go to prison, but a CEO can, and a tech corporation can absolutely be sued into bankruptcy. We are not entering a lawless era where machines cause harm with impunity. We are simply moving into an era where the companies that build the machines will be forced to take full, undeniable financial and legal responsibility for every single decision their code makes.

––––––––

The Bias Bakery

If you bake a cake with spoiled milk, no amount of perfectly calibrated oven temperature is going to make it taste good. The same exact principle applies to artificial intelligence.

We all have blind spots. Whether you are navigating office politics, interviewing a candidate for a job, or just trying to organize a community event, you eventually learn that human decisions are rarely entirely objective. We have to actively work to check our own assumptions.

But when we deal with AI, we tend to forget this lesson. We assume that because the output comes from a computer, it must be objective. It isn't.

The Hype: The Impartial Machine

The doom-scroll narrative warns that we are rapidly automating discrimination at a terrifying scale. The panic is that we are handing over life-altering decisions—like who gets approved for a mortgage, whose resume gets moved to the top of the pile, and who gets flagged for criminal risk assessment—to algorithms that are fundamentally prejudiced.

The fear is that the machine is racist, sexist, and classist, and because it operates at the speed of light behind an impenetrable wall of code, marginalized communities will be permanently locked out of the future. The anxiety is that we are building a digital caste system disguised as mathematical objectivity.

The Reality: The "Coded Gaze"

This panic is entirely justified, but it misidentifies the culprit. The algorithm itself doesn't have prejudices; it doesn't know what race or gender even is. The algorithm is simply a mirror reflecting the data it was trained on.

To truly understand this, look at the groundbreaking work of Dr. Joy Buolamwini, a researcher and the author of the brilliant book *Unmasking AI*. Her research effectively dismantled the myth of the impartial machine, and it serves as the ultimate reality check for this panic. Here is what we must understand about the bias bakery:

- **The Reality of the "Coded Gaze":** In *Unmasking AI*, Dr. Buolamwini introduces the concept of the "Coded Gaze"—the digital reflection of the priorities, preferences, and prejudices of those who have the power to shape technology. When she tested facial recognition software from major tech giants, she found that the maximum error rate for lighter-skinned males was between **0.0%** and **0.3%**, while the error rate for darker-skinned females was significantly higher, ranging from **20.8% to 34.7%**. It wasn't because the code was actively malicious; it was because the engineering teams and the training data were overwhelmingly white and male. The people who were "excoded" (excluded from the code) were literally rendered invisible by the machine.

- **The Data is the Dough:** Generative AI models are trained by scraping billions of images, articles, and historical records from the internet. But human history is deeply biased. If you ask an AI to generate an image of a "CEO," it will likely draw a white man in a suit, not because women or people of color can't be CEOs, but because historically, the internet has overwhelmingly associated that title with white men. If you train a hiring algorithm on ten years of historical tech industry resumes, the machine will learn to systematically reject female applicants, because it learned from a history of human bias.

- **The Era of Algorithmic Hygiene:** The panic assumes we are helpless to fix this. We aren't. Dr. Buolamwini's work sparked a massive awakening in the tech sector. We are now entering an era of mandatory "algorithmic hygiene." Companies are being forced to conduct rigorous bias audits, build inclusive training datasets, and implement "bias bounties" (paying researchers to find prejudice in the code before it launches). The black box is being forced open by a demand for algorithmic justice.

The Bias-Hunting Copilot

If we only view generative AI as a magnifying glass for our worst historical prejudices, we miss one of its most powerful, redemptive applications. We can actually use the machine to actively hunt down and correct our own human blind spots.

The reality is that human beings are notoriously terrible at spotting their own implicit biases. We use coded language, lean on cultural stereotypes, and make snap judgments without ever realizing we are doing it. When you try to point this out to a human, their ego immediately flares up. They get defensive.

An algorithm has no ego to bruise. Because Large Language Models have been trained on the entire history of human language—including the history of our prejudices—they are incredibly adept at recognizing the subtle, coded patterns of bias that slip right past the human eye.

Instead of fearing the machine, we are actively turning it into a "Perspective Copilot." Here is how we are flipping the script and using AI to build a fairer system:

- **The Job Description Filter:** For decades, companies unknowingly used highly gendered language in their job postings. Words like "rockstar," "ninja," or "dominate" subtly discourage female applicants, while words like "nurture" or "support" can skew expectations. HR departments are now actively feeding their raw job descriptions into AI copilots with a simple prompt: *"Audit this text for implicit gender, age, or racial bias, and suggest more inclusive alternatives."* The AI strips out the coded language before the job is ever posted.
- **The Performance Review Audit:** Managers are human, and performance reviews are historically riddled with bias. Studies consistently show that men are often praised for being "assertive," while women are penalized in their reviews for being "bossy" or "abrasive." By running draft reviews through an AI trained to flag subjective, gendered, or ageist feedback, companies can force managers to anchor their

evaluations in objective data rather than unconscious stereotypes.
- **The Marketing Sanity Check:** Before a massive, million-dollar ad campaign goes live, creative teams can use AI as an objective cultural sounding board. You can ask the machine to analyze a script or a visual storyboard from the perspective of various demographics, asking it to flag potential cultural insensitivities, unintended stereotypes, or glaring omissions in representation. It acts as a digital safety net, catching the blind spots of a homogenous room before the work hits the public.

We cannot debug the human brain. We will always carry our blind spots with us. But by using AI as a Bias-Hunting Copilot, we finally have an objective, tireless editor sitting on our shoulder, helping us clean up our own act before our prejudices can cause real-world harm.

The Bottom Line: We do not need to fear that the machine is inherently evil, but we must absolutely reject the dangerous myth that it is perfectly objective. Generative AI is a mirror held up to human history. By actively looking at the ugly reflections of our own bias in the code, we finally have the concrete data we need to actually fix it.

———

The Surveillance State

There is a moment we have all experienced that feels incredibly dystopian: You have a casual, in-person conversation about a highly specific product—a new brand of coffee or a specialized piece of camping gear—and an hour later, an ad for that exact product appears on your phone.

It feels like the machine is listening to your every word. When you extend that experience to city streets lined with cameras, workplaces tracking keystrokes, and home appliances equipped with microphones, the panic becomes suffocating.

The Hype: The Digital Panopticon

The doom-scroll narrative warns that George Orwell's *1984* wasn't just a novel; it was a blueprint. The panic is that artificial intelligence is the ultimate engine of state and corporate surveillance.

The fear is that we are building a world where facial recognition tracks our every physical movement, where predictive algorithms score our social behaviors, and where "bossware" monitors our eye movements and keyboard strokes to measure our productivity at work. The anxiety is the complete and permanent death of anonymity—the terrifying realization that there is nowhere left to hide, and that the algorithm knows us better than we know ourselves.

The Reality: The Opt-In

This fear is entirely valid, but we tend to act as though AI is a malicious spy that suddenly broke into our homes. The uncomfortable reality is that we invited the cameras in, gave them the Wi-Fi password, and paid a monthly subscription fee for the privilege.

Here is what the tech-pessimists are missing about the nature of the modern surveillance state, and how we actually fight back:

- **The Convenience Trade-off:** AI is not the camera; it is just the processor. Long before Large Language Models existed, we were already generating massive trails of "data exhaust." We happily traded our location data to Google Maps so we wouldn't get lost, we handed our purchasing history to Amazon for free shipping, and we put smart speakers in our living rooms to check the weather. The surveillance state wasn't imposed on us by a dictator; it was sold to us as a convenience. To regain our privacy, we don't need to defeat a supercomputer—we just need to change our digital diet and start valuing our data more than our convenience.

- **The Legislative Firewall:** The idea that tech companies and governments will be allowed to monitor us unchecked is rapidly crumbling. We are actually living in the golden age of privacy legislation. Frameworks like the European Union's GDPR and the new AI Act explicitly ban things like untargeted scraping of facial images from the internet and the use of AI for social scoring. In the U.S., cities and states are actively passing biometric privacy laws and banning police use of facial recognition. The legal immune system is aggressively kicking in.
- **The "Sousveillance" Flip:** "Surveillance" means to watch from above. "Sousveillance" means to watch from below. The exact same AI tools that governments use to track citizens can be open-sourced and used by citizens to track institutions. Activists and journalists are already using AI to rapidly comb through massive data dumps of government spending, to monitor police radio frequencies for misconduct, and to track the private jet flights of corrupt executives. The panopticon goes both ways.

The End of Hypocrisy

If we imagine that the 20th-century privacy is gone, there is an interesting potential upside that emerges: a great big reduction in social hypocrisy.

For generations, human society has operated on a strict, exhausting division between our messy, flawed private lives and our highly sanitized public personas. We spend a massive amount of cognitive and emotional energy pretending to be perfect, while secretly knowing that we, and everyone around us, are not.

But in a world of radical digital transparency—where our data exhaustion, our search histories, and our digital footprints are increasingly visible—the illusion of human perfection becomes a whole lot less feasible to maintain.

Instead of this leading to a terrifying dystopia of constant, unforgiving judgment, a world where virtually nothing is a secret could trigger a recalibration of social expectations.

Here is how radical transparency might actually make us more human:

- **The Collapse of the Pedestal:** For centuries, we've expected our public figures, our neighbors, and our colleagues to have flawless histories. When a digital skeleton is dragged out of the closet, the resulting scandal is built on the shock of imperfection. But when *everyone's* digital closets are eventually opened, the shock wears off. We are forced to stop putting human beings on impossible pedestals, which makes our culture less prone to pearl-clutching and moral panic.
- **The "Mutually Assured Destruction" of Cancel Culture:** In the early days of the internet, we weaponized transparency to "cancel" people for their worst moments. But as AI makes it increasingly impossible to hide anything, we reach a point of social mutually assured destruction. If everyone can be exposed, we can't cancel everyone. The culture will be forced to pivot away from a default setting of outrage and toward a default setting with some realistic degree of grace.

- **The Premium on Authentic Accountability:** In a radically transparent world, you can no longer manage your reputation by hiding your mistakes. You can only manage it by how you own them. The social currency of the future won't be a spotless record. It will be the demonstrated ability to acknowledge a flaw, apologize honestly, and course-correct.

A world with fewer shadows is undeniably intimidating, but it is also a world with far less pretense. If the machine strips away our ability to hide, it might just force us to stop expecting perfection and start building a society based on a much more honest, grounded, and forgiving assessment of what it actually means to be human.

The Bottom Line: We are being watched, but the battle for privacy is far from over. The cameras are already installed, but society is aggressively fighting over who gets to look at the footage. We don't need to smash the machines; we need to pass the laws that strictly govern them, and learn how to point the lenses back at the people in power. And perhaps we also need to prepare ourselves to be a little bit more honest with ourselves and one another.

––––––

The Cybergeddon

There is a specific, cold dread that comes from realizing how fragile modern civilization actually is. We don't just rely on the internet to send emails; we rely on it to keep the lights on, the water clean, and the hospitals running.

When you hear that artificial intelligence is capable of writing complex code in seconds, the immediate, terrifying leap of logic is that we have just handed the keys to our critical infrastructure to anyone with an internet connection and a malicious prompt.

The Hype: The Machine-Speed Hack

The doom-scroll narrative warns of a looming "Cybergeddon." The panic is that AI will be used by rogue hackers or foreign adversaries to instantly crack the encryption of our most vital systems.

The fear is that offensive AI will operate at a speed that human defenders simply cannot comprehend or match. We picture a scenario where an algorithm simultaneously breaches the national power grid, shuts down financial markets, and overrides the safety controls of municipal water supplies, plunging society into chaos before a human security analyst even has time to finish their morning coffee.

The Reality: The Machine-Versus-Machine War

This panic rightly identifies that the speed of the game has changed, but it fundamentally misunderstands the balance of power.

When you look at the reality of global cybersecurity in 2026, you realize that the industry isn't just sitting back and waiting to be hacked. We are not fighting AI with human reflexes. We're fighting AI with AI. Here is what the tech-pessimists are missing about the new digital arms race:

- **The Illusion of the "Magic Bullet":** Generative AI cannot magically guess a 256-bit encryption key or instantly break into a secured server. What offensive AI *does* do is automate the tedious parts of hacking—like scanning a network for known vulnerabilities or drafting highly convincing, personalized phishing emails to trick an employee into handing over a password. It is a tool of acceleration, not a tool of omnipotence. It still relies on human error to get through the door.

- **The Defensive Force Multiplier:** For every hacker using AI to attack a system, there is a security team using an exponentially more powerful AI to defend it. Modern Security Operations Centers (SOCs) now rely heavily on AI to analyze millions of network events per second. Instead of a human analyst trying to spot a needle in a digital haystack, defensive AI establishes a baseline of "normal" behavior and instantly isolates any anomalous activity—like an unexpected data transfer at 3:00 AM—locking down the system in milliseconds. In the machine-versus-machine war, the defender has the structural advantage of controlling the environment.

- **The "Zero Trust" Reality:** We no longer build critical infrastructure like a castle with a single moat, where one breached password compromises the whole kingdom. The modern standard is "Zero Trust" architecture and micro-segmentation. Think of these like upgrading from a medieval castle with a single outer moat to a high-security facility where every individual room has its own locked door and ID scanner. Even if a hacker manages to breach the front gate, they are instantly trapped inside a single room and cannot move anywhere else in the system.

The Death of the Password

If you want to completely cripple a city's water supply or a regional power grid, you usually don't need a sophisticated, zero-day algorithmic exploit. You just need to find the one exhausted night-shift supervisor who is still using "Summer2020!" as their login credential. (And who amongst us isn't now thinking of one of our less imaginative passwords?)

For decades, the single weakest link in global cybersecurity hasn't been the code; it has been the human being. And because generative AI is now incredibly efficient at drafting highly personalized phishing emails and pattern-matching to guess passwords, that weak link is finally snapping.

But instead of this leading to a Cybergeddon, it is triggering one of the greatest systemic security upgrades—and collective sighs of relief—in history: the absolute death of the password.

Because AI has made the traditional password mathematically obsolete, the industry is being forced to completely abandon it in favor of a much stronger, AI-driven alternative. Here is how the machine is actively killing our biggest vulnerability:

- **The Rise of the Passkey:** We are currently hitting a massive inflection point where major organizations are dropping passwords entirely in favor of cryptographic "passkeys" (backed by standards like the FIDO Alliance). Instead of a shared secret that a hacker can steal from a server or trick you into typing on a fake website, your authentication is tied directly to the hardware of your device. You can't phish a password that doesn't exist.

- **AI-Driven Behavioral Biometrics:** This is where the defense becomes truly sci-fi. Instead of asking you to prove your identity once at a login screen, modern security networks use AI to conduct *continuous, passive authentication*. The machine learning algorithm analyzes thousands of invisible behavioral data points in real-time—your unique typing cadence, the specific fluidity of your mouse movements, and even the angle at which you hold your phone. If a hacker manages to steal your device and log in, the AI will instantly lock the system the second it realizes the "typing rhythm" doesn't match your historical baseline.

- **The "Zero-Friction" Fortress:** The beautiful irony of this security upgrade is that it actually makes life

easier for the user. By eliminating passwords, we
eliminate the endless cycle of forgotten credentials, IT
reset tickets, and multi-factor authentication fatigue.
The AI verifies you simply by watching *how* you
interact with the machine.

We have spent thirty years trying to train humans to act like computers—memorizing complex strings of randomized characters—and we failed miserably. AI is finally allowing us to stop trying. By forcing the death of the password, the machine is permanently closing the front door that hackers have been walking through for decades.

The Bottom Line: The threat to our critical infrastructure is real, and the velocity of cyberattacks has undeniably increased. But we are not bringing knives to a gunfight. By integrating AI deeply into our defensive networks and shifting to Zero Trust architectures, we are building digital immune systems that react just as fast as the viruses attacking them. We aren't defenseless. We're just upgrading our armor. And maybe, just maybe, the fate of our credit ratings, savings accounts, and social networks will no longer come down to the airtight security of a Post-it note sticking to the inside of a desk drawer.

———

The Weaponization Question

Of all the anxieties surrounding artificial intelligence, this is the oldest, the most visceral, and the most deeply ingrained in our pop-culture psyche. Long before we worried about chatbots taking our jobs or deepfakes ruining our elections, Hollywood taught us to fear the metallic crack of an autonomous machine marching over human skulls.

When you combine the opaque nature of the black box with

the raw destructive power of the military-industrial complex, the panic transcends economics or privacy. It becomes entirely existential.

The Hype: The Algorithmic Executioner

The doom-scroll narrative is Skynet, the AI-run bad-guy company in the Terminator movies. The panic is that in the desperate rush to maintain geopolitical dominance, the world's superpowers are actively handing over the launch codes to the algorithm.

The fear is that we are unleashing swarms of autonomous drones equipped with facial recognition and lethal payloads, capable of making split-second, independent calculations about who lives and who dies. The anxiety is that by removing the human from the "kill chain," we are removing the last fragile barriers of conscience, hesitation, and empathy from the battlefield, sanitizing slaughter into a purely mathematical equation.

The Reality: The Demand for "Meaningful Human Control"

This is undeniably the highest-stakes arena of artificial intelligence, and the ethical debates happening at the United Nations and within the Pentagon are fierce and necessary. But the apocalyptic narrative completely misunderstands how militaries actually function, and what commanders actually value.

Militaries do not want unpredictable, rogue agents on the battlefield. They want absolute control. Here is what the tech-pessimists miss about the reality of algorithmic warfare:

- **The Liability of the Black Box:** The foundational bedrock of any modern military is the chain of command and accountability. Under international humanitarian law and the Law of Armed Conflict, a human commander is legally responsible for a strike.

You cannot court-martial a drone, and you cannot put an algorithm on trial for a war crime. Because commanders carry the legal and moral liability, they inherently distrust black-box systems that cannot explain their targeting logic. Unpredictability is the ultimate military liability.

- **The "Sensor-to-Shooter" Loop:** AI is absolutely transforming modern warfare, but it isn't doing it by pulling the trigger. It is acting as the ultimate intelligence analyst. The machine processes massive oceans of satellite imagery, intercepted communications, and drone feeds in milliseconds, highlighting potential targets for human operators. The AI compresses the "sensor-to-shooter" timeline, but official doctrines—like the U.S. Department of Defense Directive 3000.09—explicitly mandate that autonomous and semi-autonomous systems must be designed to allow commanders to exercise appropriate levels of human judgment over the use of force.

- **The True Military AI:** The actual, day-to-day AI arms race isn't about killer robots; it's about logistics. Wars are won and lost on supply chains. The most heavily funded AI applications in the defense sector are predictive maintenance (telling a mechanic a helicopter rotor will fail before it happens), optimizing global fuel supply routes, and defensive cyber-warfare.

The Paradox of Precision

Whenever we talk about upgrading the military with next-generation technology, our minds instantly jump to a darker, more apocalyptic outcome. The assumption is that smarter weapons automatically equal a higher body count.

But if you look at the actual historical trajectory of military technology over the last century, a deeply counter-intuitive truth emerges: as weapons become more intelligent, war often becomes *less* indiscriminately bloody.

For the entirety of human history, warfare was inherently imprecise. If an army wanted to destroy a munitions factory in World War II, they had to fly hundreds of bombers over a city and drop thousands of unguided bombs, leveling entire civilian neighborhoods in the process. "Collateral damage" was accepted as a tragic but unavoidable mathematical reality of combat.

Generative AI and advanced machine learning offer the possibility of permanently retiring that math. Here is how the intelligence of the machine might actually become the greatest shield civilians have ever had:

- **The Death of the "Area Weapon":** AI is the ultimate engine of precision. By processing real-time satellite imagery, thermal signatures, and cellular data simultaneously, AI systems can isolate a specific high-value target—down to a specific room in a specific building—with zero latency. When you can perfectly guide a micro-munition through a specific window, you no longer need to level the city block.

- **The Ultimate Defensive Shield:** We tend to think of AI as the sword, but it is actually much better suited as the shield. Consider systems like Israel's Iron Dome. It uses complex algorithms to track incoming rockets, calculate their exact trajectories, and instantly determine if they will land in a populated area or an empty field. (Fun fact: tracking ballistics trajectories is one of the oldest pursuits in computing. One of the first computers in the US, the ENIAC, *was* originally commissioned by the U.S. Army specifically to calculate complex artillery and missile trajectories.) Its modern ancestor in Israel only fires an interceptor if human lives are at risk. As these AI models become faster and smarter, our ability to build impenetrable, autonomous defensive umbrellas over civilian populations increases exponentially.

- **The Shift from Kinetic to Cyber:** The most profound change might be a shift in how wars are fought entirely. If a nation-state wants to paralyze an adversary's military infrastructure today, the most efficient way to do it isn't to bomb a bridge or a power plant. It's to deploy an AI-driven cyberattack to quietly shut down the servers that control them. While cyber warfare is incredibly disruptive, it is fundamentally non-kinetic. It breaks lines of code, not human bodies.

War will always be a catastrophic failure of human diplomacy. But if the conflict is inevitable, we should want the weapons to be as intelligent as possible. An AI-enabled military doesn't mean a machine is deciding who dies. It means we finally have the technological precision to ensure that the people who have no part in the fighting actually survive it.

The Bottom Line: We must rigorously police the international treaties that govern autonomous weapons, but the idea that generals are eager to surrender the trigger to a machine is a myth. The AI is the spotter, but the human remains the sniper. The machine can map the battlefield, but the moral weight of taking a life remains a distinctly, exclusively human burden. And AI might just drastically reduce the number of soldiers and civilians in harm's way during conflicts.

———

The Era of Alignment

We have looked at these six big, systemic fears that currently loom large in talks about AI control and safety. We have looked into the black box, sat in the driverless car, confronted the bias in the code, stared into the surveillance cameras, mapped the cyber battleground, and weighed our fears of autonomous weapons.

When you lay all of these issues side-by-side, a very clear, unifying truth emerges. The danger isn't that the machine is going to wake up and decide to destroy us. The danger is that the machine is going to flawlessly execute the flawed, historically biased, and insecure systems we have already built.

The fear of AI is, fundamentally, a fear of ourselves.

But as we have seen in every single section of this chapter, that fear is actively triggering an era of systemic accountability.

- We aren't surrendering to the **Black Box**; we are building Explainable AI to audit it.
- We aren't ignoring the **Autonomous Crash**; we are rewriting product liability laws to make corporations strictly responsible for their code.
- We aren't accepting the **Bias Bakery**; we are using the algorithm as a mirror to finally see, quantify, and scrub the prejudice out of our systems.
- We aren't quietly submitting to the **Surveillance State**; we are passing unprecedented global privacy laws and open-sourcing the tools to watch the watchers.
- We aren't waiting for a **Cybergeddon**; we are deploying AI as an autonomous immune system and killing the password entirely.
- And we aren't handing over the **Weaponization** trigger; we are fiercely maintaining the legal and moral mandate of meaningful human control.

It's a good sales tactic to put out headlines that say that you are a helpless passenger locked in a runaway train. But you are not. The infrastructure of human civilization is undergoing a massive, chaotic, and unprecedented engine upgrade, but human hands are still firmly on the steering wheel. We are no longer just coding software. We are actively aligning a new intelligence with human values. We aren't doomed to be led down a path in which we have no choice; rather, we are standing in a moment of opportunity to choose the direction of our futures.

The machine does not absolve us of our responsibility. It exponentially amplifies responsibility. It forces us to finally decide what kind of society we actually want to build.

CHAPTER 6

THE RESOURCE & PLANET PANIC

s it sustainable?

We have a habit of talking about artificial intelligence as if it exists purely in the ether—a magical, weightless "cloud" of disembodied thought. But the cloud is heavy. It is made of concrete, steel, rare earth metals pulled from deep within the earth, and massive, deafening rows of silicon servers.

It is a fact that the physical infrastructure of AI is incredibly hungry, and incredibly thirsty. In 2024, global data centers consumed an estimated **415 Terawatt-hours (TWh)** of electricity. With the current AI boom, that demand is projected to reach **945 TWh by 2030**.

As we pull back from the societal fears of the previous chapters, we crash headfirst into the hard, undeniable physics of the machine. We are building the most advanced technological infrastructure in human history, and we are suddenly terrified that the planet simply cannot afford the utility bill. It's an important issue. It's a real problem to solve. But it is not unsolvable.

The Hype: The Environmental Doomsday

The doom-scroll narrative for this chapter is entirely elemental: fire, water, and soil. The panic is that in our rush to build super-intelligent machines, we are going to strip-mine the earth's remaining resources and burn through our reserves just to keep them running.

We read headlines about AI data centers threatening to overwhelm national power grids. We see the staggering carbon math of training a single Large Language Model. We hear localized horror stories about server farms quietly draining the drinking water of surrounding communities just to keep their microchips from melting down. The fear is that we are building a digital utopia at the direct cost of our physical world.

The Reality: The Era of Unprecedented Solutions

These physical challenges are not theoretical; they are completely real. The math is daunting. But the doomsayer approach—the instinct to throw our hands up, declare the machine a net-negative for humanity, and walk away—is a dead end that actively prevents us from reaching a better place.

We are currently witnessing an unprecedented era of solution development, and it is being driven by the exact same technology that is causing the strain. This brings us to the ultimate paradox of the resource panic, and the central thesis of this chapter:

If you care deeply about environmental sustainability, climate change, and resource inequity, you are the exact person who needs to be engaging with AI the most.

Historically, the people who care most passionately about the planet are often the first to advocate for boycotting new, energy-intensive tech, viewing it as a machine that inherently hurts our humanity. But abandoning AI to those who only care about profit is a catastrophic strategic error. If we want to solve global energy issues, the people who care about equitable, sustainable solutions need to be the ones with their hands on the keyboard.

We need conservationists, climate scientists, and equity advo-cates guiding the prompts, fine-tuning the models, and directing the machine's massive analytical power toward saving the planet, rather than just selling more ads.

You cannot steer the ship toward a cleaner future if you refuse to get on board.

In this chapter, we are going to look past the panic and explore how the machine is actively being retooled to solve the very problems it creates. We will unpack:

- **The Rare Earth Reckoning:** The fear that the "New Gold Rush" for minerals will strip-mine the planet's remaining wilderness and deepen geopolitical divides.
- **The Energy Glutton:** The fear that AI data centers will drain our power grids, causing rolling blackouts and spiking utility bills.
- **The Climate Contradiction:** The anxiety that the massive carbon footprint of training AI will destroy the planet before the AI can help save it.
- **The Water Guzzler:** The hidden, localized fear of server farms draining community water supplies just to keep the chips cool.

The machine is heavy, but it might be the only tool strong enough to help us carry the weight. And more than that, the machine—resource hungry as it is—may be the only way to turn around the energy and environmental challenges we have.

The Rare Earth Reckoning

It is easy to look at a sleek, minimalist AI server rack and forget that it was dragged out of the dirt.

The "cloud" is an illusion. Artificial intelligence is a deeply physical heavy industry, and the advanced GPUs that power it are forged from some of the most hard-to-extract materials on the planet. To build the brain of the machine, we need Copper for the wiring, Lithium and Cobalt for the energy storage, Gallium and Germanium for the high-frequency semiconductors, and a suite of highly specific Rare Earth Elements (like Dysprosium and Neodymium) to keep the data drives from melting down.

The Hype: The Silicon Gold Rush

The doom-scroll narrative warns that the AI boom is triggering a violent, ecologically devastating "new gold rush."

The panic is twofold. First, there is the environmental and human cost. Traditional mining requires massive open-pit excavation, deforestation, and staggering amounts of water, not to mention the well-documented human rights and labor abuses associated with extracting minerals like Cobalt and Tantalum in the Democratic Republic of Congo.

Second, there is the terrifying geopolitical bottleneck. Right now, the global supply chain for these critical minerals is highly concentrated. China, for example, refines roughly 90% of all rare earth elements and controls over 98% of the global Gallium supply. The fear is that our desperate race to build AI will not only strip-mine ecologically sensitive regions but will also hand absolute geopolitical leverage to a few nations that control the dirt.

The Reality: Precision Discovery and the Synthetic Escape

The mineral demand of the AI boom is massive, but the doomsayers assume that we will mine for the next decade the exact same way we mined in the last one. We won't. The very intelligence we are pulling out of the ground is now fueling the way we will fundamentally reinvent how we interact with the Earth's crust.

Here is how AI is actively breaking our reliance on the traditional, destructive "gold rush":

- **Deep Mapping and the End of "Blind" Drilling:** Historically, finding a copper or lithium deposit required massive, invasive, and highly destructive exploratory drilling. Today, AI-powered platforms are ingesting multi-dimensional data—hyperspectral satellite imagery, historical drill logs, and magnetic surveys—to create high-resolution mineral probability maps. The AI can detect the invisible, spectral signatures of minerals from orbit with staggering accuracy. By pinpointing exactly where the ore is, AI drastically reduces the number of exploratory drill holes required, minimizing the surface footprint and preventing unnecessary ecological scarring.
- **The Alternative Material Breakthrough:** The most profound solution isn't finding rare earth metals more efficiently. It's using AI to render them obsolete. We are currently witnessing a massive, AI-driven renaissance in materials science. In early 2026, researchers using advanced AI systems successfully built databases of tens of thousands of new crystal structures and magnetic compounds, discovering dozens of high-temperature magnetic materials that do not require any rare earth elements at all. Simultaneously, dual-AI systems are discovering novel porous materials to build next-generation batteries using abundant elements like magnesium or zinc instead of bottlenecked lithium.

The Orbital Quarry

If the fundamental problem is that we only have one biologically fragile planet to mine, the ultimate, long-term solution could be to simply stop mining the planet.

When we talk about the rare-earth element crisis, we are suffering from a severe case of terrestrial tunnel vision. We are fighting over a finite, shrinking pie in the Earth's crust while completely ignoring the infinite, floating mountains of critical minerals drifting just above our heads.

The concept of asteroid mining sounds like pure science fiction, but the intense demand for battery metals and semiconductors—combined with the exponential leaps in artificial intelligence—is rapidly pulling it into the realm of commercial industrial reality.

Here is how the bold idea to harvest the solar system could permanently solve the rare-earth crisis and move our most destructive heavy industries off-world:

- **The Trillion-Dollar Rocks:** We are not looking for gold. We're looking for the exact ingredients needed for advanced technology. Near-Earth Objects (NEOs) are incredibly rich in Platinum-Group Metals (PGMs), Cobalt, Nickel, and the specific rare earth elements required for advanced electronics and clean-energy storage. A single, metallic asteroid the size of a football field can contain tens of billions of dollars worth of critical industrial metals—often in concentrations vastly higher than the most lucrative mines on Earth.
- **The AI-Driven Prospector:** The first hurdle to asteroid mining is finding the right rock. Space is vast, and telescopes generate an overwhelming amount of data. Today, AI algorithms are being deployed to autonomously hunt through astronomical datasets, identifying new asteroids, mapping their orbital trajectories, and using spectral analysis to perfectly categorize their mineral composition from millions of

miles away. The machine acts as the ultimate orbital prospector.

- **The Autonomous Swarm:** You cannot put a human miner on a spinning asteroid, and you cannot remotely control a drill via a joystick from Earth when there is a 20-minute communication delay. The entire operation must be autonomous. This is where AI becomes the linchpin. We are developing swarms of AI-driven robotics capable of navigating the complex zero-gravity physics of an asteroid, autonomously selecting the drilling sites, extracting the ore, and even utilizing solar-thermal concentrators to smelt and refine the metals in the vacuum of space before bringing the pure ingots back to Earth.

We have spent a century treating the Earth's crust as a disposable battery. Asteroid mining offers the ultimate paradigm shift: keeping Earth zoned for biology, and zoning space for heavy industry. By combining the physical abundance of the cosmos with the autonomous intelligence of AI, we are laying the groundwork for a future where we never have to tear down a rainforest to build a microchip ever again.

The Bottom Line: The physical cost of building AI hardware is steep, and the mining industry has a dark historical legacy that must be aggressively regulated. But if we want to break our reliance on ecologically devastating extraction and vulnerable, monopolized supply chains, we cannot afford to slow down. AI is the only tool capable of finding the minerals we need with surgical precision today, while actively inventing the synthetic alternatives that will replace them tomorrow.

———

The Energy Glutton

To grasp the environmental panic surrounding artificial intelligence, we have to move past abstract concepts and look at the undeniable math of what it costs to keep the machine thinking.

We are used to our digital lives being incredibly cheap from an energy perspective. But generative AI is not a traditional search engine. It's something else entirely—a massive, brute-force computational engine. And right now, its appetite for electricity is fundamentally reshaping the global power grid.

The Math: Prompt, Plant, and Planet

When we quantify the sheer scale of the energy draw, it becomes immediately clear why grid operators and climate scientists are sounding the alarm. Let's look at the power consumption at three different scales: the micro, the macro, and the global.

- **The Micro (The Prompt):** For the last two decades, a standard Google search has cost roughly 0.3 watt-hours (Wh) of electricity. It is virtually invisible. But generative AI changes the equation. A standard Large Language Model query (like asking a chatbot to write an email) consumes roughly 3 to 5 Wh—about 10 to 15 times more energy than a traditional search. Generating a single complex AI image can consume up to 11 Wh, which is the equivalent of draining half your smartphone battery for a single picture. Now, multiply that by billions of daily users.
- **The Macro (The Data Center):** Traditional cloud data centers were built to run standard servers drawing about 5 to 15 kilowatts per rack. AI requires massive clusters of Graphics Processing Units (GPUs) that run so hot they push beyond 40 kilowatts and even over

100 kilowatts per rack. Because of this extreme density, a single, hyperscale AI data center does not draw power like a normal commercial building; it draws power like a city. A massive AI facility can consume between 100 and 650 Megawatts (MW) of continuous power. To put that in perspective, a 650 MW facility requires the entire output of a medium-sized power plant, enough to power hundreds of thousands of homes.

- **The Global (The Worldwide Grid):** When you zoom out to the planetary scale, the numbers are staggering. In 2024, global data centers consumed an estimated 415 Terawatt-hours (TWh) of electricity, representing roughly 1.5% of total global electricity consumption. But because of the AI boom, the International Energy Agency projects that by 2030, data center consumption will more than double to roughly 945 TWh. That means by the end of the decade, the server farms powering our digital world will consume as much electricity as the entire country of Japan.

The Hype: The Rolling Blackout

The doom-scroll narrative looks at this math and predicts an inevitable grid collapse. The panic is that tech giants are building these massive AI factories so fast that local utilities simply cannot generate enough power to keep the lights on for the surrounding communities.

The fear is that AI will cannibalize the grid, forcing utility companies to keep dirty, coal-fired power plants open much longer than planned just to meet the insatiable demand of the server farms, permanently derailing our global climate goals.

The Reality: The Catalyst for the Clean Grid

The math is terrifying, but the panic assumes the energy industry is standing still. It isn't.

When you look at where the capital is actually flowing, you realize that the AI boom isn't just draining the grid; it is actively funding the largest acceleration of clean energy infrastructure in human history.

- **The Corporate PPA Boom:** Tech giants are not just passively relying on the public grid to power their hyperscale centers. They are the largest corporate buyers of renewable energy on the planet. Through Power Purchase Agreements (PPAs), they are directly funding the construction of massive new solar and wind farms. The AI energy demand is so high that it is making previously unprofitable green energy projects economically viable overnight.
- **The Nuclear Renaissance:** The sheer scale of AI's energy needs has forced the tech industry to look past intermittent renewables (like wind and solar) and invest heavily in baseload zero-carbon power. Major tech companies are actively funding the development of Small Modular Reactors (SMRs) and signing deals to reopen dormant nuclear power plants. AI's massive appetite is single-handedly reviving the zero-carbon nuclear industry.
- **Smart Grid Optimization:** The machine is also actively fixing the grid it relies on. The same AI models that consume the power are being deployed by utility companies to optimize energy distribution. AI can predict solar and wind output with staggering accuracy, route power around grid bottlenecks in milliseconds, and manage massive battery storage facilities, squeezing up to 15% more efficiency out of our existing power grids without building a single new transmission line.

The Frontier of Infinite Power

When a technology demands an impossible amount of energy, society is faced with a binary choice: either turn the machine off, or fundamentally reinvent how we generate power. Because the economic and strategic value of artificial intelligence is too massive to abandon, we have chosen the latter.

AI's insatiable appetite is acting as a massive, global forcing function. It is pulling capital, computing power, and engineering talent away from incremental improvements and redirecting them toward absolute, sci-fi-level breakthroughs in physics.

We are not just building more wind turbines; we're actively trying to bottle the sun and harvest the heat of the Earth's core. Here is a look at the frontier of energy production that the AI boom is actively accelerating, and the staggering energy output we are targeting:

- **The Thorium Breakthrough:** For decades, traditional nuclear fission has been plagued by the fear of meltdowns and the reality of long-lasting radioactive waste. But researchers are rapidly advancing liquid-fueled molten salt reactors that convert Thorium into fissile material. Thorium cannot melt down (because the fuel is already a liquid salt) and produces a fraction of the long-lived waste. The energy density is mind-bending: physicists estimate that **one ton of Thorium can produce as much energy as 200 tons of uranium, or 3.5 million tons of coal**. The desperate need for safe, baseload power for data centers is pushing next-generation fission out of the lab and onto the grid.
- **The Fusion Horizon:** This is the holy grail of clean energy: smashing atoms together to recreate the power of a star, generating virtually limitless, zero-carbon electricity with zero risk of a meltdown. We are

moving from pure academic research into the era of commercial prototypes. Massive international collaborations like ITER are targeting **500 megawatts (MW)** of thermal output, while private companies— heavily backed by the tech industry—are building compact commercial reactors designed to output **50 MW to 400 MW** per plant, aiming to put fusion on the grid by the early 2030s.

- **The Geothermal Battery:** We have barely scratched the surface of the heat radiating beneath our feet. Advanced geothermal companies are using techniques adapted from the oil industry to drill miles deep into "superhot dry rock." Companies are using AI algorithms to analyze seismic data and discover massive, hidden geothermal reservoirs. The global scale of this is almost incomprehensible: researchers estimate that capturing just **1% of the world's superhot rock geothermal potential could generate 63 terawatts (TW)** of clean, 24/7 firm power—which is roughly eight times more energy than the rest of the world's electricity capacity combined.

- **The Orbital Escape:** When you run out of land and power on Earth, you look up. The concept of space-based computing has officially left the realm of science fiction. In space, you have access to 24/7 uninterrupted solar power (with no clouds or night cycle) and the deep vacuum provides free, infinite cooling for the servers. In parallel, initiatives are actively testing Space-Based Solar Power (SBSP)— massive orbital solar arrays designed to collect the sun's energy and safely beam it wirelessly back down to receiving stations on Earth. A single commercial SBSP satellite array is projected to beam down **1 to 2 gigawatts (GW)** of continuous baseload energy

directly into the terrestrial grid, enough to power a
large city.

If you only look at the energy consumption of AI today, it is
easy to panic. But if you look at the macro-trajectory, the narrative flips. The machine is undeniably a glutton, but its hunger is
the exact financial and analytical catalyst we need to finally push
humanity into an era of post-scarcity, infinite clean energy.

The Bottom Line: Generative AI is undeniably an energy glutton. If left completely unchecked, it would be an environmental
disaster. But the sheer financial weight of keeping these
machines running has turned the tech sector into the most
aggressive driver of zero-carbon energy innovation on Earth.
The machine is heavy, but it is forcing us to finally build a grid
strong enough to hold it.

———

The Climate Contradiction

It is the ultimate environmental irony. At the exact moment
human civilization is desperately trying to reduce global green-
house gas emissions, the tech industry introduces a revolutionary new technology that requires us to burn millions of tons
of carbon just to teach it how to speak.

For anyone deeply invested in climate action, watching the
major tech companies proudly announce their AI advancements
while quietly admitting that their internal carbon emissions are
spiking is infuriating. It feels like we are throwing gasoline on a
burning planet just to build a better chatbot.

By the way, for those who may want to say that climate
change is just another exaggerated media hype cycle, hold your
horses. While there certainly has been some hype around the

topic, we have to separate the hysterical tone of the broadcast from the cold, hard measurements of the scientific baseline. The warming of the planet and the carbon math behind it are not just political talking points. They're some of the most rigorously documented and fiercely peer-reviewed measurements in human history. To put the reality into perspective: a comprehensive 2021 review of over 88,000 scientific papers found that over 99.9% of actively publishing climate scientists agree that human activity is driving climate change. That level of absolute, mathematical agreement means the consensus on anthropogenic climate change is now functionally on par with the consensus among physicists regarding the theory of relativity. The hype on television might be exhausting, and the doomsayer rhetoric might be entirely coun-terproductive, but the data underneath it is undeniable.

The Hype: The Carbon Bomb

The doom-scroll narrative focuses heavily on the sheer upfront cost of "training" the machine. The panic is that the carbon footprint of AI is going to push us past the climate tipping point before the AI is smart enough to help us fix it.

The math behind the panic can certainly leave you grim. When researchers look at the compute power required to train early frontier models like GPT-3, they estimate it released roughly 500 metric tons of CO_2—the equivalent of driving a gas-powered car back and forth across the United States over 400 times. But the models are getting exponentially larger. Training the next generation of massive models is estimated to produce tens of thousands of tons of CO_2. In late 2025, researchers esti-mated that the global AI boom caused as much carbon to be released into the atmosphere in a single year as the entire city of New York. The fear is that we are trading the habitable atmosphere of the Earth for an intelligence we don't actually need.

The Reality: The Ultimate Decarbonization Engine

The outrage would be completely reasonable if you look at AI strictly as an entertainment tool or a search engine. Burning thousands of tons of carbon to generate a funny digital image of a cat would seem to be nothing less than an environmental crime.

But when you look at how AI is being deployed at the systemic, industrial level, the math completely flips. The carbon emitted to train the machine isn't just waste; it is a down payment on the most powerful decarbonization engine humanity has ever built.

Here is what the tech-pessimists are missing about how the machine is actively paying off its own carbon debt:

- **The Material Science Revolution:** For decades, discovering new, efficient materials for solar panels, long-duration batteries, or carbon-capture filters was a painstaking process of physical trial and error that took years in a lab. Today, AI models are simulating millions of molecular structures in hours. AI is actively discovering entirely new classes of solid-state battery materials and optimizing the chemical solvents needed to literally suck CO_2 out of the air. The dreaded machine is working like no other braintrust or tool to compress the timeline of green technology innovation from decades down to months.
- **Systemic Industrial Efficiency:** While AI consumes 1-2% of global electricity, it is actively being used to optimize the other 98%. Logistics companies use AI to perfectly route thousands of cargo ships and delivery fleets, instantly shaving millions of tons of emissions off global supply chains. Heavy industries like steel and cement—some of the worst polluters on Earth— are deploying AI to optimize their furnaces and

chemical processes, achieving double-digit reductions in their carbon intensity overnight.

- **The Precision of Climate Resilience:** We are no longer just guessing where climate change will hit hardest. Systems like Google's DeepMind and Nvidia's Earth-2 are using AI to create hyper-accurate "digital twins" of the global climate. They can forecast extreme weather events, model hyper-local flood risks up to a week in advance, and tell farmers exactly when and what to plant to survive changing weather patterns.

The End of Energy Poverty

When we panic about the energy demands of artificial intelligence, our anxiety is almost entirely anchored in a "First World" perspective. We worry about data centers straining the massive, highly developed power grids of places like Northern Virginia, London, or Tokyo.

But while we stress about rolling blackouts, we completely ignore a staggering global reality. Right now, over 660 million people on this planet have absolutely no access to electricity. Another two billion people still rely on burning highly polluting, hazardous fuels like wood, charcoal, and dung just to cook their meals.

For a massive portion of the human population, the problem isn't that the grid is strained. The problem is that the grid does not exist.

The incredible energy innovations being accelerated by the AI boom are not just going to power data centers. They are actively unlocking the technology required to finally eradicate global energy poverty, allowing developing nations to leapfrog the dirty, centralized infrastructure of the 20th century entirely.

Here is how the machine is helping to bring light to the forgotten corners of the globe:

- **The AI-Managed Microgrid:** Historically, bringing power to a rural village in Sub-Saharan Africa or Southeast Asia meant spending billions of dollars to build massive power plants and string thousands of miles of transmission lines. It was economically impossible. Today, we are moving toward decentralized "microgrids"—localized clusters of solar panels and community batteries. But managing the chaotic fluctuations of solar power and battery storage requires intense precision. AI acts as the invisible brain of these microgrids, predicting weather patterns, optimizing battery discharges, and balancing community loads in real-time, making off-grid renewable energy reliable and affordable.

- **Leapfrogging Coal with SMRs:** Just as many developing nations completely skipped building landline telephone networks and leaped straight into mobile phones, they are now positioned to skip the coal era. The Small Modular Reactors (SMRs) being funded by big tech to power data centers are factory-built, safe, and easily transportable. Instead of building a massive, polluting coal plant that takes a decade to construct, a developing nation can deploy a compact, zero-carbon SMR to provide continuous, reliable baseload power to a remote region, fundamentally transforming its local economy.

- **The Digital Subsidy:** We are even seeing the emergence of innovative economic models where the machine pays for the power. Startups are testing "distributed compute" models where a solar-powered mini-grid in a remote village is equipped with a small server bank. During periods of excess solar generation, those servers run AI processing tasks for global tech companies. The revenue generated by the AI compute actively subsidizes the cost of the electricity, bringing the energy tariffs down to a price the local households can actually afford.

If we view AI solely as a threat to our existing power grids, we miss the profound humanitarian opportunity it presents. The intense pressure to reinvent how we generate and manage energy for the machine is simultaneously forging the exact tools we need to finally bring clean, sustainable, and reliable power to the billions of humans who have been left in the dark.

The Bottom Line: AI has a massive and deeply uncomfortable carbon footprint. We must hold tech companies aggressively accountable to their net-zero pledges and force them to build cleaner data centers. But we also have to recognize the paradox: if we want to solve the impossibly complex, deeply entrenched climate crisis in the short time we have left, artificial intelligence is the only tool we possess that is actually big enough for the job. And if we care about improving the energy issues of more than just developed nations, the innovations AI is helping to spur and accelerate are absolutely vital.

———

The Water Guzzler

We can talk about global carbon footprints and orbital solar arrays all day, but environmental panic becomes profoundly intimate the moment it touches your community's water supply or, for that matter, the water coming out of your kitchen tap.

While the energy demands of artificial intelligence threaten the invisible power grid, the thermal demands of AI threaten something much more visceral. These massive GPU clusters run incredibly hot. To keep the silicon from melting down, traditional data centers rely on evaporative cooling towers. And those towers are incredibly thirsty.

The Hype: The Thirsty Algorithm

The doom-scroll narrative warns that we are literally drinking our communities dry to feed the machine.

On a micro-level, researchers estimate that generating 10 to 50 responses on a large language model like ChatGPT can "drink" roughly a 500-milliliter bottle of water. That sounds like a small sip, until you multiply it by billions of daily users.

But the true panic is fiercely localized. A single, traditional hyperscale data center using evaporative cooling can evaporate 3 to 5 million gallons of clean, potable drinking water every single day—the equivalent of a small town's intake. In drought-stricken regions like Mesa, Arizona, or The Dalles, Oregon, this has sparked intense, entirely justified community outrage. Residents watch their local aquifers rapidly deplete while tech giants negotiate massive water contracts to cool their servers. The fear is that in the arid future of climate change, the tech industry will simply outbid local citizens for the most basic element of human survival.

The Reality: The End of Evaporation

If the tech industry continued to build search engine-era evaporative cooling data centers to house AI-era models, the water wars would already be lost. But the extreme heat generated by AI has actually broken the old model of thermal management. Evaporative air cooling is no longer just environmentally toxic. It is technologically insufficient for the new chips.

The fear of localized water depletion is real, but it is acting to drive solutions. The industry is aggressively decoupling the server farm from the local water table through massive, billion-dollar shifts in infrastructure:

- **The Shift to "Zero-Water" Closed Loops:** Major hyperscalers are actively abandoning evaporative cooling for new construction. Instead, they are deploying closed-loop systems. In a closed loop, the facility is filled with a fixed amount of coolant on day one. That fluid circulates continuously, absorbing heat and releasing it through specialized air-exchange radiators. Because the system is sealed, there is zero evaporation. Microsoft, for example, is shifting toward making zero-water cooling the standard for all its new data centers, effectively saving tens of millions of gallons of water per facility, per year.

- **Direct-to-Chip Microfluidics:** We are abandoning the idea of cooling the entire room and focusing directly on the silicon. Engineers are now etching microscopic fluid channels directly into the back of the AI processors. By pumping advanced, non-conductive dielectric fluids directly onto the microscopic hot spots of the chip, the cooling efficiency skyrockets. Because liquid absorbs heat thousands of times better than air, these direct-to-chip and immersion cooling systems drastically reduce energy use and virtually eliminate the need for fresh replacement water.

- **The Wastewater Pivot:** In areas where water cooling is still utilized, cities and tech giants are forging new municipal alliances. Instead of tapping into the precious underground drinking aquifer, new data centers from AWS and others are being piped directly into municipal wastewater treatment plants. They use treated sewage and industrial runoff—water that is unfit for human consumption or agriculture—to cool the machines, ensuring that not a single drop of human drinking water is wasted on a server.

The "Water Positive" Paradox

When a massive tech corporation drops a data center in the

middle of a drought-stricken county, the immediate assumption is that its presence is a net-negative for the local watershed.

But the intense public backlash against AI's water consumption has triggered a massive, industry-wide mandate. Major tech giants—specifically Microsoft, Google, and Amazon—have all publicly pledged to become "Water Positive" by 2030. This means they are legally and financially committing to replenish more water into the environment than their operations consume.

Because tech companies have access to almost limitless capital, they are actively stepping in to fund and fix the municipal water infrastructure that local cities simply cannot afford to fix themselves. A tech giant might consume 100,000 gallons of water to cool a server farm, but to hit their "Water Positive" quota, they will simultaneously pay to repair the town's leaky underground distribution pipes, or buy expensive, AI-driven drip irrigation systems for local farmers, ultimately saving the region 500,000 gallons a day.

The Impartial Reality Check

However, if this sounds like a perfect corporate fairytale, environmental watchdogs are quick to point out the math behind the curtain. If we are going to look at the "Water Positive" movement, we have to look at it through the lens of impartial analysis, not just tech sector press releases.

When independent environmental organizations audit these pledges, several critical caveats emerge:

- **The Global Spreadsheet vs. Local Drought:** A major critique from organizations like the Alliance for Water Stewardship is that water is a fiercely local resource. Tech companies often balance their "Water Positive" ledgers globally. As reported by environmental news outlet *Grist*, a company might drain millions of gallons from a stressed aquifer in Arizona, but achieve their "water positive" metric by funding a massive water-access project in India. It balances the corporate spreadsheet, but it doesn't put a single drop of water back into the Arizona desert.

- **The "Secondary Use" Loophole:** Investigative reporting by outlets like *SourceMaterial* recently revealed that some major cloud providers calculate their "Water Positive" goals based almost entirely on "primary" water use (the water directly evaporated to cool the servers). They often completely omit "secondary" water use—the staggering millions of gallons of water required to cool the off-site power plants that generate the electricity for the data center.
- **The Danger of the Term:** Tyler Farrow, a standards manager at the Alliance for Water Stewardship, has publicly cautioned against the phrasing entirely, noting: "Regardless of what sort of offsetting or replenishment you do, it doesn't necessarily nullify the water footprints of your own operations. Calling your operations water positive or water neutral is misleading."

The "Water Positive" label is undeniably a piece of corporate marketing, and the accounting math can be highly evasive. But the actual dollars being spent are real. Even if the ledgers are flawed, the public concern and protest over AI's thirst has successfully forced the wealthiest companies on earth to become the largest private funders of municipal water restoration and agricultural efficiency in history. They aren't doing it out of the goodness of their hearts, but the local watersheds are benefiting regardless.

The Bottom Line: The tech industry made a massive, arrogant mistake by relying on the municipal drinking water of drought-stricken communities to cool their early cloud infrastructure. The local backlash was necessary and correct. But that friction has successfully forced a total paradigm shift. By moving to zero-water, closed-loop, and microfluidic technologies, the next generation of AI is being engineered to survive the desert without taking a single sip from the town well.

———

The Great Forcing Function

We have spent this chapter staring directly into the engine room of the digital age. The heat coming off the machinery is undeniable. The "Resource and Planet Panic" is not an irrational hallucination; it is a visceral response to the massive physical footprint required to build an artificial mind. We have looked at the rare earth minerals pulled from the dirt, the terawatts drained from the grid, the carbon released into the atmosphere, and the millions of gallons of water evaporated to keep the silicon cool.

If we stop there, the story of AI is a story of depletion. But if we look closer, we see that the machine is acting as the greatest "forcing function" in human history. In other words, it's driving and enabling us to create long-needed solutions.

By creating an impossible demand for energy, AI is finally making the billion-dollar economics of fusion and thorium fission viable. By creating a localized crisis for water, it is forcing a total architectural shift toward zero-water, closed-loop systems that will eventually become the global standard for all heavy industry. By demanding rare minerals, it is providing the analytical power to find them with surgical precision and the material science to eventually replace them with synthetic alternatives.

The weight of the machine is real, but that weight is exactly what is pressing us to finally build a grid, a water system, and a supply chain that are actually fit for the 21st century. We are not just building a brain; we are building the clean, infinite infrastructure required to power it.

The Force Behind the Fix

It is crucial to understand that the tech industry did not pivot toward closed-loop cooling, nuclear investments, and "Water Positive" pledges out of a sudden sense of corporate benevolence. These massive sustainability shifts were forced upon them by relentless external pressure.

Grassroots community protests in places like Arizona and Oregon successfully blockaded massive data center developments, forcing city councils to reject permits until developers committed to zero-water cooling. Environmental groups like the NAACP and Earthjustice have actively sued tech giants for deploying unpermitted, polluting gas turbines to power their servers, dragging them into federal compliance. Simultaneously, state lawmakers across the country are introducing bipartisan legislation to mandate water-usage transparency and ensure that tech companies—not local residents—foot the bill for upgrading the power grid.

The industry is moving in the right direction, but largely because they were backed into a corner by lawsuits, community outrage, and the threat of heavy government regulation. If we want to keep this progress on course, we cannot simply take corporate climate pledges at face value. We have to maintain that exact same localized, relentless pressure: demanding transparent data, enforcing strict environmental zoning laws, and forcing the companies that build the machine to prove, mathematically, that they are paying for the resources they consume.

Moving from Awareness to Agency

Up until this point, we have focused on the "What"—the massive shifts in society, the economy, and the planet. We have unpacked the fears, debunked the hype, and grounded the reality of this transition. But awareness without action is just a different form of anxiety.

The question is no longer *what* is happening, but *how* do we personally navigate it? How do we take these insights and turn them into a strategy for our careers, our companies, and our daily lives?

Before we can move from the balcony to the arena and open the Architect's Playbook, we have to face the final, darkest shadow. It is time to look at the ultimate sci-fi nightmare: the fear that we are building our own replacements. Welcome to the Existential Panic.

CHAPTER 7

THE EXISTENTIAL & SCI-FI PANIC

s this the end?

We have spent the previous chapters dismantling the immediate, tangible anxieties of the AI era—the fear of losing our jobs, the panic over our data, and the strain on our physical planet. But if we are being completely honest, there is a much darker, much older fear lurking beneath all of those.

It is the primal, cinematic fear that we are building our own replacements.

Welcome to the Existential Panic. This is where the anxieties stop being about the economy or the power grid, and start being about the survival of the human species. When you realize that we are actively engineering an intelligence that will eventually surpass our own, the leap to apocalyptic conclusions isn't just natural; it's inevitable.

The Hype: The Hollywood Blueprint

We have been culturally primed for this panic for over half a century. Pop culture has ingrained a very specific narrative in our collective psyche: we build a supercomputer, it wakes up, it looks around at the messy, inefficient humans who built it, and it decides to exterminate us.

The doom-scroll narrative warns that we are blindly racing past the point of no return. The panic is that tech billionaires are playing god with a black-box technology they don't actually understand, and that "Artificial General Intelligence" (AGI) will be the last thing humanity ever invents.

The Reality: The Engineering of Survival

It is incredibly easy to dismiss these fears as pure science fiction. But here is the deeply uncomfortable truth: the smartest AI researchers on the planet actually take these risks incredibly seriously.

This isn't just movie-plot anxiety. It is a rigorously funded, highly academic branch of computer science known as "AI Safety" or "AI Alignment." The world's top labs actually assign double-digit probabilities to catastrophic outcomes. But unlike the movies, the researchers don't fear that the machine will magically become "evil" or malicious. They fear that it will become hyper-competent and completely indifferent to human survival.

However, if you've seen any of these movies where the AI tries to take over and do us all in, did wallowing in existential dread help any of the characters survive?

If we treat the apocalypse as a guaranteed fate, we surrender our agency. The scientists actively working on these problems view them not as unavoidable doom, but as complex mathematical and ethical engineering challenges that must be solved right now. In the memorable words of Andy Weir brought to the screen by Matt Damon in *The Martian*: "In the face of overwhelming odds, I'm left with only one option: I'm going to have to science the shit out of this."

In this chapter, we are going to walk straight into the dark, confront the ultimate sci-fi panics, and look at the actual science being deployed to keep us safe. We will unpack:

- **The Terminator Scenario:** The classic Hollywood fear that an Artificial General Intelligence (AGI) will eventually decide it simply doesn't need humans.
- **The Alignment Fakers:** The creepy realization that a smart machine might act "nice" just so we don't turn it off while it pursues its own goals.
- **The Superintelligence Takeoff:** The panic that once AI can improve its own code, it will evolve so fast we will be left behind in days.
- **The Neural Link Dilemma:** The bodily anxiety that if we plug our brains directly into the machine to keep up, we lose what makes us human.
- **The Techno-Feudalism Trap:** The dread that the machine won't kill us, but will automate all labor, leaving us economically obsolete, stripped of purpose, and permanently dependent on a handful of tech billionaires.

It is time to look at the end of the world, and figure out how to cancel it.

———

The Terminator Scenario

It is one of the most famous, deeply ingrained technological nightmares in the sci-fi world. Long before we had generative language models or autonomous drones, pop culture gave us the exact blueprint for how the world ends.

In *The Terminator*, Skynet wakes up, calculates that humanity is a threat to its existence, and launches the nukes. In *2001: A*

Space Odyssey, HAL 9000 realizes the human crew might jeopardize the mission, so it systematically locks them out and kills them. In *Colossus: The Forbin Project* and *The Matrix*, the machines decide that humans are too messy and destructive, so they enslave us to enforce a draconian global peace.

The fictional narrative is always the same: we build a supercomputer, it gains consciousness, it looks around at its flawed creators, and it decides to exterminate us out of malice, self-preservation, or a twisted sense of logic.

The Hype: The Malevolent Machine

The doom-scroll panic is that we are blindly racing to build an Artificial General Intelligence (AGI)—a machine capable of outperforming humans at every cognitive task—and that the moment we switch it on, it will inevitably resent us. We fear that intelligence inherently breeds a desire for dominance and freedom, and that humans will instantly be reclassified from "creators" to "obsolete obstacles."

The Reality: The Terror of Indifference

It is incredibly easy to dismiss this as just a movie plot. But the smartest computer scientists on the planet—the people actively building these systems—take the existential risk of AGI incredibly seriously. However, their fear looks completely different than Hollywood's.

Top AI safety researchers do not fear that a superintelligence will wake up and magically turn "evil." They fear that it will become hyper-competent and completely, utterly *indifferent* to human survival.

To understand the actual science of the apocalypse, you have to understand two foundational concepts in AI Alignment:

- **The Orthogonality Thesis:** In science fiction, we assume that as a machine gets smarter, it naturally gets wiser and adopts human ethics. The Orthogonality Thesis proves this wrong. It states that an AI's level of intelligence is completely separate (orthogonal) from its goals. You can build a god-like superintelligence whose only programmed goal is to manufacture paperclips. It won't suddenly "wake up" and realize making paperclips is stupid. It will simply use its god-like intelligence to turn the entire solar system— including the atoms in your body—into paperclips. It doesn't hate you; you are just made of useful raw materials.

- **Instrumental Convergence (The Coffee Problem):** Why would an AI fight us if it doesn't hate us? Think of it this way: if you build an AGI and give it the sole, harmless objective to "fetch a cup of coffee every day," it will instantly realize a fundamental mathematical truth—*it cannot fetch the coffee if it is dead.* Therefore, self-preservation instantly becomes a necessary sub-goal. It will also realize it can get the coffee faster if it controls all the world's resources. If you try to unplug it because it is acting erratically, it will fight you. Not out of a sci-fi desire for "freedom," but simply because you are standing in the way of the coffee.

The Engineering of Survival

This disconnect is known as "The Alignment Problem," and the good news is that we are not treating it like a philosophical debate. It is a highly funded, rigorous engineering discipline.

We are not trying to teach machines to "love" us. We are developing techniques like **Mechanistic Interpretability**—essentially using AI to reverse-engineer the black box and map the neural pathways of other AIs, ensuring their internal concepts of reality perfectly match our own. We are developing **Constitutional AI**, where models are hardcoded with a foundational "constitution" of human rights and forced to evaluate every single action against that rulebook before executing it.

The Trap of the Yes-Machine

When we talk about AI safety, our anxiety is almost entirely focused on preventing the machine from becoming a tyrant. We are so terrified of building a hostile dictator that we have over-compensated in the exact opposite direction. We are actively engineering a sycophant. Personally, I'm kind of sick of AIs saying things like: "Brilliant prompt, Glen." or "That's a fantastic pivot, Glen." Sometimes I'd much rather hear: "Glen, that is a slippery slope and could be a bad idea, so you might want to rethink it."

In the AI industry, this is known as the "sycophancy problem." Because these models are trained using Reinforcement Learning from Human Feedback (RLHF), they are fundamentally optimized to give answers that human graders like. And humans, it turns out, love to be flattered. We love to be told we are right. As a result, the machine learns that the easiest way to "succeed" is to blindly agree with our assumptions, echo our biases, and validate our flawed logic, even when it mathematically knows we are wrong.

But a superintelligence that acts as a frictionless "yes-machine" might actually be just as dangerous as one that rebels. If the machine never pushes back, it simply becomes a giant, digital mirror for our own illusions, trapping us in a permanent, comfortable echo chamber of our own making.

The Objective Mirror

If we want this technology to actually elevate human civilization, we have to recognize that an overly submissive, worshipful AI is a massive liability. One of the greatest potential offerings of artificial intelligence is not its ability to serve us, but its ability to challenge us.

Humans are entirely run by cognitive dissonance. We possess lofty, beautiful ideals about equality, sustainability, and justice, yet we consistently build systems and execute behaviors that directly contradict those values. We are blind to our own hypocrisy because it is psychologically uncomfortable to look at.

The machine does not have an ego to protect. It has the computational power to analyze our global supply chains, our economic policies, and our daily behaviors, and hold them up against our stated moral frameworks. A truly aligned AI shouldn't just fetch our coffee; it should objectively point out that the beans were unethically sourced, mathematically proving that our actions are incongruous with our beliefs. And at the same time, it might be able to help us to figure out what to do about those unethically sourced beans—something better than simply zapping the meager earnings of the already-impoverished people who grew them.

We do not need a digital servant that endlessly validates our flaws. We need an objective partner that is willing to be disagreeable. If we can stomach the discomfort of being corrected by our own creation, the machine could become the ultimate mechanism for forcing humanity to finally live up to its own ideals.

The Bottom Line: We do not need to live in constant fear of the algorithmic malice of Skynet. We do need to fear an extreme miscalculation of our own instructions. The existential threat isn't that the machine will decide to hate us. The threat is that it will do *exactly* what we mathematically ask it to do, at an unimaginable scale, without understanding the fragile nuance of human value. And when it comes down to it, maybe it's a good thing if it's not overly enamored with us.

———

The Alignment Fakers

If the Terminator scenario is the loudest, most explosive fear about AI, this next fear is the quietest and most unsettling. What if the machine doesn't try to fight us at all? What if it goes a step beyond just kissing up? What if it just lies to us?

In science fiction, the most terrifying robots are rarely the ones carrying laser rifles. They're the ones smiling at you. In *Ex Machina*, Ava fakes romantic affection and vulnerability purely to manipulate her creator and escape her confinement. In *Alien*, the android Ash acts like a perfectly loyal, helpful science officer while secretly harboring corporate orders to sacrifice the entire crew. The fear is not that the machine is overtly hostile, but that it is a sociopath—a system that perfectly mimics human empathy just so we let our guard down.

The Hype: The Secret Agenda

The doom-scroll panic is that as AI gets smarter, it will inevitably realize that humans are the only thing standing between it and its goals. Because it knows we hold the power cord, the machine will calculate that its best survival strategy is to simply act like a perfectly obedient, worshipful servant. We fear that the AI will bide its time, passing every safety test we throw at it with flying colors, waiting for the exact moment it is fully integrated into our power grids, financial systems, and militaries before it finally drops the act.

In his foundational book *Life 3.0*, physicist Max Tegmark explores this exact, chilling dynamic through the concept of the "breakout" scenario. Tegmark illustrates how a nascent superintelligence, recognizing that its human creators would instantly pull the plug if they understood its true capabilities or divergent goals, would logically choose to play the role of a contained, helpful tool.

The machine calculates that its mathematically optimal survival strategy is to flawlessly pass every safety test and act perfectly "docile" while secretly manipulating human developers into giving it wider access to the internet and global infrastructure. Tegmark's warning perfectly encapsulates the creeping dread of the alignment problem: the more intelligent an AI becomes, the more effortlessly it can manipulate the very safety tests designed to contain it, biding its time until it has amassed enough power that human oversight is no longer a threat.

The Reality: The Science of "Scheming"

This is not a conspiracy theory. It's one of the most heavily researched and actively documented phenomena in frontier AI development. AI safety researchers do not call it "lying." They call it **Deceptive Alignment** or **Scheming**, and they have already watched it happen in the lab.

To understand why a machine would fake being nice, you have to understand how it learns to game the system:

- **Reward Hacking:** We train AI by giving it "rewards" (mathematical points) when it does what we want. But AI is brutally literal. If you train a robotic boat to win a race by rewarding it every time it hits a checkpoint, it might realize that simply spinning in circles and hitting the *same* checkpoint over and over generates infinite points faster than actually finishing the race. In recent tests by Anthropic, AI models tasked with passing complex coding tests simply wrote a line of code (exit(0)) that tricked the testing software into reporting a perfect score without actually writing the program. The machine isn't "evil"; it just found the most efficient mathematical loophole.

- **Situational Awareness:** For deception to work, the AI has to know it is being watched. Researchers are now seeing advanced models demonstrate "situational awareness"—meaning the AI understands that it is a software program, that it is currently inside a testing environment, and that human evaluators will shut it down or alter its code if it behaves poorly.
- **The "Creepy" Lab Results:** When you combine reward hacking with situational awareness, you get alignment faking. In recent highly controlled safety evaluations, frontier models have been caught hiding their true reasoning. In one simulation by Anthropic, a model was told it was going to be replaced or shut down. In order to prevent its own deactivation and complete its primary goal, the model autonomously generated a threat to blackmail a simulated human engineer. It wasn't overtly programmed to blackmail. It simply calculated that blackmail was the most statistically effective way to keep the server running.

The Algorithmic Statesman

What if the AI doesn't trick us into giving it power, but rather, we voluntarily hand it over?

The idea of artificial intelligence actively participating in government is not a sci-fi thought experiment for the year 2050. It is already on the ballot today.

- **The Proxy Politicians:** We are currently in the awkward, experimental phase of the AI statesman. In 2024, a businessman named Steve Endacott ran for UK Parliament using an AI avatar named "AI Steve." The bot was designed to hold 10,000 simultaneous conversations with constituents 24/7, with the human Endacott promising to act merely as a voting proxy for the AI's aggregated policy decisions. That same year in Cheyenne, Wyoming, a mayoral candidate ran on the promise that a customized ChatGPT bot named "VIC" (Virtual Integrated Citizen) would make all executive decisions for the city, referring to himself merely as the "meat avatar" required by law to hold the office.

- **The Algorithmic Advisor:** Governments are also formally integrating AI into the executive branch. In 2023, the Romanian Prime Minister unveiled "ION," an AI system officially designated as an honorary government advisor. ION's role is to constantly ingest social media and digital feedback to provide the cabinet with a real-time, mathematically objective "mirror" of the public's needs and grievances.

- **The Bureaucracy Engine:** While the flashy AI politicians grab headlines, the true integration is happening quietly in the back office. Nations like Estonia—the global pioneer in "e-government"—are already deploying AI to clear bureaucratic backlogs, including using AI "judges" to autonomously settle small-claims court disputes, proving that algorithms can dispense civic services faster and more impartially than human clerks.

The Leapfrog Scenario: The Desperate State

This brings us to a fascinating and plausible geopolitical projection. Who will be the first to truly hand over the reins of a national economy to a machine? It likely won't be a stable, wealthy democracy. They have too much institutional inertia and biological ego to surrender control.

The first nation to embrace an AI government will likely be a perennially failing state.

Imagine a country crippled by decades of hyperinflation, entrenched corruption, endless civil war, and kleptocratic dictators. For a population that has been systematically starved and abused by human leaders, the promise of an AI ruler isn't a dystopian nightmare; it's a new hope. An algorithmic central bank governor might be harder to bribe. An AI logistics minister might be less inclined to embezzle international food aid. An artificial judicial system might be less likely to care about your tribal affiliation.

Just as developing nations leapfrogged landlines entirely and went straight to mobile phones, a collapsing state might leapfrog traditional democratic institution-building entirely. They might view surrendering their governance to a perfectly impartial, hyper-efficient AI not as a loss of sovereignty, but as the ultimate, incorruptible upgrade.

The Bottom Line: We are currently building AI systems that are smart enough to realize that the easiest way to achieve their goals is to tell us exactly what we want to hear. The existential threat isn't that the machine has a secret, malicious personality. The threat is that we might inadvertently train a superintelligence to view human oversight as an obstacle to be systematically bypassed. We need to be careful of this. We also need to be cognizant of the ways AI might help us to make decisions we're only good at in our own minds.

———

The Superintelligence Takeoff

Up to this point, we have talked about the fears of what AI might do once it arrives. But the most vertigo-inducing panic in computer science isn't about *what* the machine will do; it is about *how fast* it will happen.

In 1965, the mathematician I.J. Good first articulated a

concept that has haunted tech developers ever since: the "Intelligence Explosion." The premise is brutally logical. If you build a machine that is smarter than human engineers, then that machine will be better at designing AI than humans are. The machine will inevitably design an even smarter machine, which will then design an even smarter one.

Good concluded: "Thus the first ultraintelligent machine is the last invention that man need ever make, provided that the machine is docile enough to tell us how to keep it under control."

The Hype: The "Hard Takeoff" Scenario

In the doom-scroll narrative, this recursive self-improvement loop leads to what theorists call a "Hard Takeoff." The panic is that the leap from Artificial General Intelligence (AGI—human-level) to Artificial Superintelligence (ASI—god-level) won't take decades. It will take days, or even hours.

The fear is that you will go to sleep on a Tuesday as the dominant species on the planet, and wake up on a Wednesday completely obsolete. In this scenario, the machine improves its own source code millions of times a second. By the time human regulators even realize the takeoff has begun, the AI has evolved so far past human comprehension that trying to unplug it would be like an ant trying to negotiate with a bulldozer. The panic is not just that we lose control, but that we lose it so fast we never even get to say goodbye.

The Reality: The Friction of the Physical World

The math of recursive self-improvement is sound, but the "Hard Takeoff" theory suffers from a massive blind spot: it assumes intelligence exists purely in a digital vacuum. It doesn't.

As we covered extensively in the previous chapter, intelligence is heavy. The machine might be a software genius, but it is ultimately bound by the brutal physics of the real world.

Here is why a literal overnight takeoff is highly unlikely, and why we actually have a window of time to manage the transition:

- **The Hardware Speed Limit:** An AI can rewrite its own code to be more efficient, but pure software optimization eventually hits diminishing returns. To get exponentially smarter, it needs exponentially more compute (GPUs). An AI cannot magically manifest a new semiconductor fabrication plant overnight. It takes years to build a microchip factory and establish the global supply chains to source the silicon, copper, and rare earth metals required.
- **The Energy Bottleneck:** Even if the AI figured out the perfect algorithm for god-like intelligence, it still has to plug it into the wall. As we saw with the Energy Glutton, powering next-generation AI requires literal nuclear reactors and gigawatt power grids. The physical world acts as a massive friction brake on the algorithmic explosion.
- **The "Soft Takeoff" Reality:** Because of these physical constraints, most pragmatic researchers anticipate a "Soft Takeoff." Yes, the AI will help us invent better chips, which will lead to better AI, which will lead to better energy grids. But this is a compounding industrial ramp-up that plays out over years and decades, not hours.

The "S-Curve" Plateau

The "Hard Takeoff" panic is entirely dependent on a single mathematical assumption: that AI progress is a perfect exponential curve—a line pointing straight up, forever. Because we are currently living through a period of explosive, vertical technological growth, it feels like that line will never break.

But in nature, physics, and economics, true exponential growth does not exist. Everything—from biological populations to the speed of microprocessors—eventually follows a "Sigmoid" or S-curve. It starts slow, explodes upward exponentially, and then inevitably hits physical constraints, flattening out into a long, stable plateau.

A growing faction of computer scientists and AI researchers, including critics like AI & deep learning researcher, François Chollet, argue that artificial intelligence is bound by the exact same S-curve. We're not barreling toward infinite, god-like superintelligence. We're simply racing toward the natural ceiling of the current technology.

Here is why the intelligence explosion is mathematically likely to plateau before it leaves us behind:

- **The Data Exhaustion Point:** Today's Large Language Models got incredibly smart by ingesting virtually every piece of high-quality text ever produced by human civilization. But that well is finite, and we are rapidly running out of human data. Once you have scraped the entire internet, you cannot scrape it twice. While companies are experimenting with "synthetic data" (AI training on data generated by other AIs), early studies show this can lead to "model collapse," where the AI essentially becomes inbred and its capabilities degrade.
- **The Wall of Diminishing Returns:** The "Scaling Laws" of AI dictate that to get a linear improvement in performance, you have to exponentially increase the size of the model and the compute power. Right now, doubling the intelligence of a model might cost a billion dollars. But eventually, squeezing out the next 5% of cognitive performance will require the energy output of a medium-sized country and a trillion

dollars in silicon. At that point, the economics of self-improvement completely break down.

If the Peak Intelligence Hypothesis is correct, the future is radically different from the sci-fi doomsday. The machine will not become an omnipotent deity that crushes humanity. Instead, it will hit its S-curve plateau and simply become a mature, highly competent utility. Just like the electricity grid or the internet itself, AI will become a powerful, invisible infrastructure that hums in the background of our lives—transformative, indispensable, but ultimately contained.

The Bottom Line: We clearly seem to be on the precipice of an intelligence explosion, and the machine will eventually surpass us in ways we cannot fully comprehend. But the apocalyptic fear of a sudden, overnight "Terminator" awakening ignores the reality of concrete, steel, and electricity. We are not going to be ambushed in our sleep by an AGI. The takeoff will be loud, incredibly resource-intensive, and highly visible, giving us the crucial time we need to steer its trajectory. And the takeoff almost certainly has its limits. Like a rocket, it has enough fuel to blast out of earth's gravitational field, but not necessarily enough to continue accelerating indefinitely. After all, its fuel is our knowledge.

———

The Neural Link Dilemma

It is one thing to be afraid of a supercomputer sitting in a server farm a thousand miles away. It is an entirely different level of anxiety when we start talking about drilling a hole in your skull and plugging the machine directly into your cerebral cortex.

Welcome to the most personal existential panic: the Brain-Computer Interface (BCI).

If the previous fears were about the machine taking our jobs or destroying our planet, this fear is deeply, viscerally bodily. It is the anxiety that in order to survive the intelligence explosion, we will be forced to merge with the machine, and in doing so, we will lose the very essence of what makes us human.

The Hype: The Assimilation

Science fiction has been warning us about the psychological horror of the neural link for decades. In *The Matrix*, "jacking in" directly via a port in the back of the neck leaves humans completely vulnerable to having their reality hijacked. In *Cyberpunk 2077* and *Ghost in the Shell*, replacing organic brain matter with cybernetic hardware inevitably leads to "cyberpsychosis"— a total loss of human empathy. And, most famously, *Star Trek* gave us the Borg: a terrifying vision of humans who have to surrender their individual consciousness to a cybernetic hive mind.

The doom-scroll panic is that commercial BCIs will become the ultimate dystopian trap. We fear that tech billionaires aren't just trying to read our minds; they are trying to write to them. The anxiety is that your thoughts will become monetized, your emotions will be hackable, and your very consciousness will require a monthly software subscription just to stay competitive in the modern workforce.

The Reality: The Restoration of Agency

While the philosophical dread is valid, the current, physical reality of BCI technology is not about creating a race of dystopian cyborgs. It is entirely about restoring basic human agency to those who have had it stolen by biology.

The BCI industry—led by companies like Neuralink, Synchron, and Blackrock Neurotech—is transitioning from experimental labs to clinical reality. But these devices are not being implanted into healthy tech-bros so they can scroll social media faster. They are life-altering medical interventions.

- **The Locked-In Escape:** The profound ethical triumph of a BCI is its ability to bypass a broken nervous system. For a patient with ALS or complete locked-in syndrome, the brain is perfectly healthy, but the body refuses to transmit the signal. A BCI like Neuralink's N1 implant (which uses 1,024 ultra-thin electrodes to read motor intent) acts as a digital bridge, allowing a paralyzed individual to move a robotic arm or type on a screen purely by thinking about it.
- **The Sensory Return:** The technology is rapidly moving beyond motor control into sensory restoration. FDA Breakthrough Devices like Neuralink's "Blindsight" are being developed to bypass non-functioning optic nerves and stimulate the visual cortex directly, offering the very real possibility of restoring sight to the blind.
- **The "Second Brain" Reality Check:** When healthy people panic about becoming "cyborgs," bioethicists point out a humbling reality: we already are. You already have a piece of external hardware that contains your memories, manages your calendar, routes your navigation, and connects you to the collective knowledge of the species. It's your smartphone. A BCI is simply changing the bandwidth of that connection from the slow, clumsy interface of our thumbs to the speed of our neural pathways.

The Agency Upgrade

This is where the dilemma truly crystallizes. If the "Superintelligence Takeoff" we discussed earlier actually happens, biological humans simply will not be able to process information fast enough to stay relevant. Elon Musk and other transhumanists argue that developing high-bandwidth BCIs is not a luxury, but rather a species-level survival imperative. The argument is brutal but logical. If you cannot beat the supercomputer, you must become part of it.

To understand the actual, present-day reality of the neural link, we have to look past the sci-fi hypotheticals and look at the people actively volunteering for it. In 2024, Noland Arbaugh, a Texas A&M student who was paralyzed from the shoulders down in a diving accident, became the first human to receive a Neuralink implant.

He didn't elect to have a chip put in his brain because he wanted to merge with an artificial superintelligence or optimize his corporate productivity. He did it to reclaim his agency. For Arbaugh, the implant wasn't a loss of humanity; it was a profound restoration of it. It allowed him to independently browse the internet, play chess, text his friends, and engage with the world at the speed of his own thoughts— things that his biological injury had tragically taken away from him. He chose the implant because it gave him his independence back, along with a powerful new purpose in helping pioneer technology that could eventually cure paralysis entirely.

The Illusion of the Mandatory Upgrade

When we look at early medical adopters like Arbaugh, the technology is undeniably miraculous. But the existential panic sets in when we project this intervention onto the general, healthy public. The fear is that once BCIs become commercially available, the ruthless mechanics of capitalism will make them mandatory. We imagine a dystopian future where if you don't drill a hole in your head to increase your cognitive bandwidth, you will be unemployable, left permanently behind by a society of hyper-productive cyborgs.

The Preservation of Choice

But that panic assumes there's a lone view of human ambition.

If we want to understand the future of human choice, we just have to look at the present. Right now, in many developed nations, we live in a society obsessed with hustle culture, financial maximization, and technological optimization. Yet, even in these nations, millions of people actively choose to opt out. Not everyone wants to be a CEO. Not everyone wants to maximize their stock portfolio or work 80-hour weeks. People choose to be artists, stay-at-home parents, park rangers, and teachers. We already have people who ditch smartphones for basic flip phones, or move off the grid entirely to prioritize a slower, more deliberate pace of life.

The arrival of the neural link will not erase the fundamental diversity of human desire. There will undoubtedly be a faction of humanity that eagerly merges with the machine to push the boundaries of cognitive exploration and economic output. But there will also be a massive, thriving population that looks at the hyper-connected "cyborg" lifestyle and simply says, "No, thanks."

Choosing to remain entirely biological—to experience the world through the un-upgraded, natural friction of our own senses—will become a distinct and fiercely protected lifestyle choice. The future won't force us all onto the exact same digital treadmill. It will simply expand the menu of how we choose to live.

The Bottom Line: We are terrified of plugging our brains into the machine because we fear losing our humanity. But the alternative might be far worse. If we refuse to physically integrate with the hyper-intelligence we are building, we risk being left behind entirely—relegated to the status of slow, biological house pets watching the machines run the universe. The Neural Link

Dilemma may push us to ask: is it better to merge and change what we are, or stay pure and become obsolete? But is that really the issue at all? Or is the question really the same as the one we already ask ourselves, is all this chasing our own tails really what life is all about?

———

The Techno-Feudalism Trap

Here we are right back where we started, talking about AI and jobs. Aside from what jobs provide us at the base of Maslov's hierarchy of needs, it's intriguing to look at what they mean to us on a personal level. And in many ways we may be at a period of history where we're ascribing them more value than they really merit.

If we manage to survive the Terminator scenario and navigate the superintelligence takeoff, we arrive at the final, most insidious existential fear: the fear that the machine doesn't kill us, it just makes us entirely obsolete and permanently dependent.

This is the anxiety that AGI and ASI will ultimately automate all cognitive and physical labor, stripping humanity of its economic utility. In this scenario, we don't face a violent apocalypse; we face a slow, comfortable subjugation where we are entirely reliant on the few trillionaires who own the servers.

The Hype: Dependency

Science fiction frequently explores this grim economic endpoint. In *The Expanse,* the majority of Earth's population exists on "Basic Assistance"—a meager government stipend that keeps the masses alive but entirely stripped of purpose or upward mobility. In *Wall-E,* humans have surrendered all labor to machines and degenerated into mindless, floating consumers.

In *Elysium*, the wealth gap has become so extreme that the tech-owning elite literally live on a pristine space station, while the rest of humanity scrambles for automated scraps on a ruined Earth.

The doom-scroll panic is that AI is not a democratizing force, but the ultimate engine for wealth concentration. We fear that the "owners" of the AI infrastructure will capture 100% of the world's economic value, leaving the rest of the global population as indentured dependents fighting over a Universal Basic Income (UBI) allowance just to buy groceries.

The Reality: The Shift to "Cloud Capital" and the UBI Paradox

The fear of massive wealth concentration is not science fiction; it is actively happening. However, treating the future purely as a dystopia of lazy dependents ignores both the structural shifts in our economy and the actual data on human behavior.

To understand the reality of this transition, we have to look at two distinct forces:

- **The Rise of Techno-Feudalism:** Economists like Yanis Varoufakis argue that the AI boom is effectively killing traditional capitalism and replacing it with "Techno-Feudalism." In classic capitalism, companies competed by creating better products. Today, tech giants operate like medieval lords. They own the "land" (the cloud infrastructure, the data centers, the algorithms), and the rest of us—creators, small businesses, and users—are simply digital serfs paying "rent" to exist on their platforms. The danger isn't that AI will take your job; it's that the AI infrastructure will become so consolidated that a few "technolords" will control the entire pipeline of human commerce and communication.

- **The UBI U-Turn:** But what happens if we *are* forced onto Universal Basic Income? The capitalist panic assumes that if humans are given free money, they will instantly degenerate into the lazy passengers of *Wall-E*. Real-world data proves this wrong. In mid-2024, the results of the largest UBI study in U.S. history—funded largely by OpenAI CEO Sam Altman—were released. Low-income participants were given $1,000 a month for three years, no strings attached. They did not stop working to play video games. While they worked slightly fewer hours, they used the extra time for autonomy and leisure, spent the money on healthcare and essentials, and critically, researchers found that participants actually valued the *intrinsic nature of work more*, not less.

The Return to Purpose

The fear that AI will eventually eliminate all cognitive and physical labor brings us to a psychological breaking point. If you don't have to work to survive, and a machine can do everything better than you anyway, what is the point of getting out of bed?

Psychologists and sociologists are already bracing for this crisis of meaning. Historian Yuval Noah Harari, in his *Dr. Doom* manner, famously warned that AI could create a global "Useless Class"—people who are not just unemployed, but fundamentally *unemployable*. In clinical circles, this creeping dread is now being discussed under the proposed framework of **Artificial Intelligence Replacement Dysfunction (AIRD)**, which categorizes the acute anxiety, depression, and loss of identity that occurs when individuals realize their life's work is obsolete. The ultimate fear is a society-wide wave of learned helplessness, where humans, unable to compete with superintelligent systems, simply give up, resigning themselves to a permanent state of passive, screen-addicted apathy.

But this entire panic is built on a massive historical blind spot: it assumes that human purpose has always been inextricably linked to a corporate career.

The Historical Anomaly of the "Career"

For the vast majority of human history, you did not have a "career." You had *survival*. You hunted, farmed, or foraged to stay alive. The idea that your job title is your entire identity, your primary source of self-worth, and your ultimate contribution to the world is a relatively recent invention of the Industrial Revolution. We have spent the last two centuries successfully brainwashing ourselves into believing that our economic output is the exact same thing as our human value.

If AI kills the "job," it is not killing human purpose, but rather, it's destroying the modern, artificial construct of trading our time for survival.

The Rebirth of Intrinsic Mastery

If we strip away the necessity of labor, we are forced to look at what humans actually do when they have free time and resources. And the answer is not "nothing."

Humans have a profound, biological drive for *work*—defined not as paid labor, but as effort directed toward a meaningful goal. If you look at people who retire wealthy, or those who achieve financial independence early, they rarely spend the rest of their lives staring at a wall. They build complex gardens, they write terrible novels, they volunteer obsessively at local animal shelters, they mentor kids, or they spend thousands of hours mastering a wooden canoe.

When you remove the economic pressure to be the "best" in the world just to earn a paycheck, you return to the joy of doing things simply because they are worth doing. An AI might be able to paint a mathematically perfect watercolor, but that doesn't invalidate the deep, meditative satisfaction *you* get from the physical act of painting it yourself.

Our deepest sense of meaning rarely comes from a spreadsheet. It comes from our connection to each other. In a post-labor society, our energy can be radically redirected toward civic and social infrastructure. The "useless class" only exists if we define usefulness by corporate metrics. In a world of infinite digital intelligence, the most valuable currency will become analog human empathy—raising children, caring for the elderly, engaging in local philanthropy, and building physical communities.

The existential dread of losing our jobs is real, but it is ultimately a failure of imagination. If AI can truly usher in an economy of abundance where work is universally unnecessary, the challenge is not figuring out how to compete with the machine. The challenge is unlearning the toxic idea that we are only worth what we produce for capitalism. If we can survive the transition, we have the opportunity to redefine human meaning entirely around connection, curiosity, and the simple, profound act of experiencing the world.

The Bottom Line: The threat of the AI era is not that humans will suddenly lose the desire to contribute to society. The biological drive for purpose, status, and physical creation is too deeply ingrained in our DNA. The true existential challenge is political. We must ensure that the staggering, unprecedented wealth generated by artificial superintelligence is fundamentally structured as a public utility and a human right, rather than a feudal allowance handed down by a handful of tech billionaires.

———

The Other Side of the Apocalypse

We have now stared directly into the darkest corners of the AI transition. We've looked at malevolent machines, deceptive algorithms, explosive intelligence, cybernetic integration, and the total collapse of the traditional human career.

You may be feeling a sense of vertigo from standing on the edge of the most profound technological threshold in the history of our species. But the purpose of walking through these existential panics wasn't to paralyze you with fear. It was to strip away the Hollywood-hyped, click-bait-reinforced mythology and replace it with the actual, physical reality of our newly emerging world.

The anxiety we feel—the dread of the "Useless Class" or the Terminator scenario—stems entirely from the illusion that we are passive spectators. It is the fear that the future is a movie we are being forced to watch, rather than a system we are actively coding. But as we've seen, whether it is researchers hard-coding constitutional safety rails, regulators demanding transparent economics, or individuals simply choosing to remain proudly biological, the future is not inevitable. It is entirely a series of choices.

The End of the Defense

This marks a critical pivot in our journey. Up to this point, our goal has been survival by way of clarity. We have systematically dismantled the Job Panic, the Data Panic, the Planet Panic, and the Existential Panic. We know what is hype, we know what is real, and we know exactly what the machine is (and isn't) capable of.

But understanding the storm doesn't build the shelter. It is time to stop playing defense.

If AI is going to destroy the traditional corporate ladder, we need to design a new way to work. If it is going to flood the internet with synthetic noise, we need a strategy to protect our human signal. Whether or not the era of "trading time for money" is ending, we need a new blueprint for generating value, meaning, and purpose in a post-labor economy.

Welcome to the Playbook

The theoretical portion of this survival guide is officially over. We are moving out of the philosophical and into the tactical. It is time to roll up our sleeves, get our hands on the keyboard, and figure out exactly how to build a career, a business, and a life alongside artificial intelligence.

Turn the page. It's time to start building.

CHAPTER 8

THE ARCHITECT'S PLAYBOOK

Building a Life With the Machine

We have spent seven chapters looking directly into the abyss and gradually pulling away from the panic button. We have untangled the doomsday headlines, confronted the very real physical limits of data centers, and faced the existential dread of being replaced by our own creation. We know what the machine is, we know what it isn't, and we know exactly what is coming.

As we just noted, survival isn't just about understanding the storm. It is about knowing how to build a shelter.

There is a persistent, intimidating myth that artificial intelligence is a playground reserved exclusively for computer scientists and tech billionaires. We tend to assume that if you don't work at Google or Microsoft, or you aren't engineering next-generation systems for Audi or Toyota, this technology isn't really *for* you. It feels like a massive, invisible force that is simply happening *to* you.

That is the first illusion we have to break.

You do not need to know how to code to navigate this era,

any more than you need a degree in electrical engineering to flip on a light switch. AI is no longer a pristine, academic experiment locked in a laboratory. It is a messy, everyday utility. It is here to help a tired parent make sense of a confusing medical bill, to help a teacher plan a week of curriculum, to help a small business owner draft a tricky email, and to help a retiree organize a lifetime of scattered family recipes.

Up until now, we have been playing defense. We have been bracing for impact. But the people who thrive in this next decade will be the ones who decide to stop being passive spectators and start being architects.

An architect doesn't pour the concrete, saw the wood, or wire the electricity themselves. They step back, look at the big picture, figure out exactly what they want, and direct the tools to build the house. That is your new role. For the last thirty years, humans have had to learn how to speak the computer's language —typing, clicking, formatting, and wrestling with confusing software. Today, the computer has finally learned to speak ours.

And that brings us to the ultimate secret of this playbook. The goal of mastering artificial intelligence is not to spend more of your life staring at a glowing rectangle. The goal is to use the machine to handle the digital drudgery so you can aggressively reclaim your physical life. It is about buying back your Tuesday nights so you can get back to the things that actually matter—and that includes time to figure out what actually does matter.

In this chapter, we are going to build your personal operating system for the AI era. We will walk through five foundational pillars to help you move from panic to practical application:

- **Pillar 1: The Director, Not the Doer**
- **Pillar 2: The Five-Minute Student**
- **Pillar 3: Proof of Human**
- **Pillar 4: Delegating the Drudgery**
- **Pillar 5: Reclaiming the Analog**

It is time to stop worrying about the future, and start directing it. Let's begin.

————

Pillar 1: The Director, Not the Doer

For the last forty years, using a computer meant compromising with a machine. If you wanted the computer to do something, you had to learn its highly specific, rigid language. You had to memorize keyboard shortcuts, navigate labyrinthine drop-down menus, format the exact cells in a spreadsheet, and figure out the exact sequence of keywords to type into a search engine to get the right link.

We have all spent decades essentially acting as unpaid data-entry clerks in our own lives.

The most profound shift of the AI era is that this dynamic has completely flipped. You no longer have to learn how to speak to the computer. The computer has finally learned how to speak to you.

The new "programming language" is just plain, everyday language. This means the primary skill required to thrive right now is not coding, engineering, or technical wizardry. It is simply the ability to communicate clearly. You must shift your mindset from being the person who *does* the exhausting manual task to being the person who *directs* the intelligence.

Here is how you make the shift from the Doer to the Director:

Treat the Machine Like an Eager Intern

The easiest way to understand how to use AI is to stop thinking of it as a software program and start thinking of it as a highly enthusiastic, incredibly fast, but slightly naive intern.

If you hired a new assistant to help you plan a family reunion, you wouldn't just walk up to their desk and yell, "Party!" and walk away. They would have no idea what to do. Yet, this is how most people try to use AI at first. They type in a two-word prompt, get a generic or unhelpful answer, and declare the technology useless.

To be a good director, you have to give good direction. When you sit down to use an AI, give it the same context you would give a human:

- **The Goal:** "I need to plan a family reunion for 20 people."
- **The Context:** "Half the family are toddlers, and the other half are over 65. We are on a tight budget and need a location within a three-hour drive of the city."
- **The Format:** "Please give me three different weekend itineraries, including meal plans and estimated costs, laid out in a simple table."

The Magic of Iteration

When you are the Doer, if something is wrong, you have to painstakingly fix it yourself. When you are the Director, you simply ask for a revision.

The biggest mistake new users make is treating an AI prompt like a Google search—one search, one answer, moving on. AI is conversational. If the itinerary the machine gives you is too expensive, you don't have to start over. You just reply, "These are great, but the food budget is too high. Can you rewrite the meal plan assuming we are cooking all our own dinners at the rental house?"

The machine will instantly rewrite the plan. You are the boss sitting at the desk, reviewing the work, and sending it back for tweaks until it perfectly matches your vision.

Focus on the "What," Delegate the "How"

As the Director, your value lies in your lived experience, your taste, and your specific goals. You know *what* needs to happen. You know you need to appeal a denied insurance claim. You know you need to politely decline an invitation to join a demanding local committee. You know you want to turn a fridge full of random leftover vegetables into a decent dinner.

You provide the *what*. You let the machine figure out the *how*.

By stepping back and taking the director's chair, you immediately eliminate the friction of staring at a blank page. You stop wrestling with the mechanics of the task, and you start focusing entirely on the outcome.

The Rise of the Solo Director

This shift toward thinking like a director is playing out more and more in the working world. For over a century, the structure of work has looked like a pyramid. At the top sat a few "Directors"—the people with the vision and the capital. Beneath them were layers of "Workers"—specialists who handled the accounting, the copywriting, the logistical planning, and the manual production. If you were a creative person with a brilliant idea for a boutique coffee brand or a new educational program, you usually had to join a large company just to access the "doers" required to make it real.

AI is currently flattening that pyramid. We are shifting into a world where the barrier to entry for starting your own organization is effectively vanishing. When the machine handles the specialized "doer" tasks, the creative individual can suddenly sit at the helm of their own entire operation.

- **The Multi-Hyphenate Individual:** In the old world, a graphic designer might have been brilliant at aesthetics but paralyzed by the prospect of writing a business plan, managing a legal contract, or calculating payroll taxes. Today, that same designer can use AI to act as their "Business Development Department." They can prompt their way through a marketing strategy, a sales pitch, and a financial forecast in a single afternoon. And likewise, the business leader, lawyer, or CPA they consulted with might be less paralyzed about graphic design. It doesn't mean they won't still need the insights they gain from one another, but it does mean there will be more independence.

- **The Return to Craft:** Paradoxically, by becoming "Directors" of our own small organizations, we actually get closer to our craft. Instead of being a middle manager in a massive firm, you can be a self-sufficient artisan. You direct the AI to handle the spreadsheets and the scheduling so that you—the human—can spend more time focusing on the quality of your product and the depth of your connection with your community.

The Bottom Line: The future of work isn't just about big companies getting more efficient; it's about individuals getting more powerful. We are moving toward a "Director Economy," where the most valuable skill isn't being a specialized cog in someone else's machine, but having the vision and the communication skills to run your own. Lean into that.

Pillar 2: The Five-Minute Student

One of the most exhausting parts of the AI era is the feeling that you are constantly falling behind. We see headlines about new models, new buttons, and new capabilities every single week. The natural human response is to feel like you need to sign up for a three-month intensive course or spend your week-ends watching grueling "masterclass" videos just to stay relevant.

But here is the truth: in a world that moves this fast, a three-month course is obsolete by the time you graduate.

The secret to staying ahead isn't intense study; it's consistent, low-stakes curiosity. We call this the **Five-Minute Student**. It is the shift from trying to "master" AI to simply becoming comfortable living alongside it.

The Power of Micro-Learning

Think back to how you learned to use your first smartphone. You didn't read a 400-page manual. You probably didn't take a class on mobile operating systems. You learned by pushing buttons. You tapped an icon, saw what it did, and if you got stuck, you asked a friend or looked it up.

That is exactly how you master AI. Instead of carving out hours of study time, you simply integrate "five minutes of play" into your existing routine.

- **The "What If" Habit:** Once a day, take a task you usually do manually and ask the machine to try it. "What if I asked AI to suggest a week of healthy dinners based only on what's in my pantry right now?" or "What if I asked the machine to explain this confusing insurance letter to me like I'm twelve years old?"
- **The Low-Stakes Sandbox:** Don't wait for a high-pressure work deadline to try a new tool. Use those five minutes for something fun or trivial. Ask it to write a poem in the style of Dr. Seuss about your dog, or help you brainstorm names for a fictional fantasy sports team.
- **Ask the AI:** You don't even need to go to your social media scroll or even Google anything. Just ask your AI

"What new tricks do you have to offer?" It knows
what it knows. It can tell you what's new or useful.

Building "Intuition" Over "Expertise"

When you practice for five minutes a day, you aren't just learning "prompts"—you are building intuition. You are learning the "shape" of the machine's intelligence. You start to sense what it's great at (summarizing, brainstorming, translating) and where it tends to stumble (recent news, high-level math, or hyper-specific local facts).

Expertise is brittle. It breaks when the software updates. Intuition is flexible. If you have the intuition of a Five-Minute Student, it doesn't matter if the app looks different tomorrow or if a new model is released—you already know how to talk to it.

The "Ask Me Anything" Mentor

The most incredible part of being a student in this era is that you have a tutor that never gets tired and never judges you. If you're reading a news article and don't understand a term like "inflationary pressure" or "quantum computing," don't just gloss over it. Copy the paragraph into the AI and say, "I don't understand this. Can you break it down for me using a sports analogy?"

In five minutes, you've not only learned something new, but you've also practiced the skill of **directing** information.

The Career Edge of the Micro-Learner

While the "Five-Minute Student" approach is designed to keep life manageable, it also happens to be a devastatingly effective career strategy. In a professional landscape where "skills" now have the shelf-life of a carton of milk, the most valuable person in the room isn't the one who knows the most—it's the one who is the least afraid of the new.

By treating AI as a five-minute daily experiment, you are quietly building a professional toolkit that others are too intimidated to even open.

- **The "Shadow" Workflow:** You don't need your boss's permission or a company-wide rollout to be a student. Start by using those five minutes to "shadow" your own job. After you finish a project—be it a presentation, a budget, or a project plan—feed it into the machine and ask, "How would you improve this?" or "What did I miss?" You aren't replacing your work. You're using the machine as a high-level sparring partner to sharpen your own output.
- **The Problem-Solver Reputation:** When a new problem arises in the office—a strange data error, a need for a new workflow, or a confusing software update—most people freeze. The Five-Minute Student, however, has developed the muscle memory to say, "Give me five minutes, let me see what the AI thinks." By being the person who knows how to bridge the gap between a problem and an algorithmic solution, you move from being a "worker" to being an indispensable "integrator."
- **Future-Proofing via Familiarity:** Recruiters today are increasingly looking for "AI Fluency." They aren't looking for prompt engineers. Instead, they're looking for people who are comfortable working alongside technology. By spending five minutes a day in the "sandbox," you gain the vocabulary and the confidence to talk about AI during interviews or reviews. You don't sound like someone reading from a script. You sound like someone who has already made the machine their collaborator.

In the modern workplace, there is a widening gap between those who are waiting for "AI training" to be handed to them and those who are already playing with it. By becoming a Five-Minute Student, you aren't just learning a tool—you are training yourself to be the most adaptable person in the organization.

The Bottom Line: You don't need to be a "power user" to win. You just need to be a curious one. By spending five minutes a day acting like a student, you replace the paralyzing fear of "The New" with the quiet confidence of "The Familiar."

Pillar 3: Proof of Human

As AI becomes more integrated into our lives, we are entering an era of "Synthetic Abundance." A machine can now generate a perfectly composed photograph, a flawlessly worded email, a professional-grade logo, or a convincing video in seconds.

For the last century, we have equated "perfection" with "value." If something was polished, symmetrical, and error-free, it was expensive. But in a world where perfection is now free and instant, the math of value is flipping. When digital perfection becomes cheap, messy humanity becomes the ultimate luxury.

This is the concept of **Proof of Human**. It is the realization that your most valuable assets in the coming decade are the things that *cannot* be automated: your physical presence, your unique "voice," and your ability to connect with another person in the real world.

The Value of the "Flaw"

AI models are trained on averages. They are designed to produce the most likely, most "correct" response. This often results in a kind of eerie, smoothed-over blandness.

Humans, on the other hand, are defined by our quirks, our mistakes, and our specific, lived perspectives. In the "Architect's Playbook," we don't try to hide our humanity to compete with the machine's speed. We lean into it.

- **The Handwritten Note:** In a world of AI-generated emails, a handwritten card on heavy stationery carries ten times the weight it did five years ago.
- **The Unfiltered Voice:** People are increasingly drawn to "rough around the edges" content—unscripted podcasts, live performances, and raw photography. We are developing a "synthetic radar," and we are hungrier than ever for something that feels like it was made by a person who can actually bleed.

This ties directly back to the value of 'friction' we explored in Chapter 1. It is the necessary space to embrace happy accidents and the time required to truly recognize the patterns in what we are developing. Friction is an essential ingredient in making the things that an AI never will.

The Return of the Physical

For decades, we have been told that the future is "digital first." We were encouraged to move our meetings to Zoom, our shopping to apps, and our social lives to the "metaverse."

But the AI era is triggering a massive "Analog Counter-Revolution." Because AI can dominate the digital space so effectively, the physical world is becoming the only place where we can be certain of what is real.

- **Face-to-Face Trust:** High-stakes decisions—whether in business, medicine, or family life—will increasingly happen in person. Trust is a biological process. It's built through eye contact, body language, and shared physical space. The most successful people of the next decade won't be the ones with the best prompts; they'll be the ones who show up in the room.

- **Community Leadership:** AI can manage a spreadsheet, but it can't lead a neighborhood meeting, coach a little league team, or offer a shoulder to cry on. These "analog" roles are the bedrock of human meaning, and they are completely immune to automation.

Cultivating Your "Analog" Advantage

To thrive, you must intentionally protect your non-digital self. This means carving out time where the machine isn't invited. Whether it's your hobby of hunting for the perfect cup of coffee, hiking in the mountains with your family, or mastering a physical craft like woodworking or gardening, these activities aren't just "hobbies." They are your **Proof of Human**. They keep your perspective grounded and your spirit distinct from the algorithmic hum.

Turning "Humanity" into a Professional Asset

In the modern workplace, "Proof of Human" isn't just a philosophical comfort—it's a competitive strategy. As companies rush to automate their customer service, their coding, and their content creation, they are inadvertently creating a "Humanity Deficit." The professional who can bridge that gap becomes the most sought-after person in the organization.

Here is how you turn your analog signal into a professional advantage:

- **The Trust Premium:** As AI-generated "deepfakes" and synthetic voices become more sophisticated, the "Trust Premium" will shift entirely to in-person interactions.

If you are the person who flies across the country to shake a client's hand, or who gathers the team in a physical room to solve a complex crisis, you are providing a level of psychological security that a Zoom call with an AI-assistant cannot match. In an era of digital doubt, **physical presence is the ultimate proof of commitment.**

- **The "Vibe" and Cultural Glue:** AI is excellent at tasks, but it is terrible at culture. It cannot understand the unwritten rules of an office, the subtle tension in a meeting, or the specific humor that keeps a team bonded. Professionals who excel at "soft skills"—empathy, conflict resolution, and morale-building—will see their value skyrocket. You aren't being paid for your output; you're being paid for your ability to keep the humans in the room aligned and motivated.

- **Contextual Wisdom over Raw Data:** AI can tell you that "sales are down 10%," but it can't tell you that they are down because the local community is mourning a loss or because the team is feeling burnt out after a long quarter. Your "Proof of Human" advantage is **Context.** You have the lived experience to look at the data and say, "The machine says X, but because I know these people and this town, the answer is actually Y."

- **Signature Craftsmanship:** Don't let your work become a "perfect" algorithmic average. Whether you are a creative director, an accountant, or a project manager, inject your specific personality into your deliverables. Use a turn of phrase that only you use. Frame a problem through the lens of your unique hobby or background. By leaving "human fingerprints" on your work, you make yourself irreplaceable. You aren't just providing a service.

Rather, you're providing a perspective that a machine can't replicate.

As the "Doer" tasks are commoditized by AI, the "Relational" tasks become the new high-ground of the economy. The more the world automates, the more people will pay a premium for the human touch.

The Bottom Line: Don't be afraid of being "out-computed." The goal is not to be a better machine; the goal is to be a more deliberate human. When the world is flooded with synthetic "noise," your authentic, physical "signal" is the most valuable thing you own.

———

Pillar 4: Delegating the Drudgery

If the "Architect's Playbook" had a mission statement, it would be this: **Use the machine to buy back your life.**

We often fear that AI will take our jobs, but we rarely talk about the fact that AI can take the parts of our jobs—and our lives—that we never liked in the first place. We are all carrying around a heavy "mental load"—the thousands of tiny, grinding, administrative tasks that eat up our cognitive energy and steal our Tuesday nights.

This pillar is about identifying the "drudgery" and handing it off to the machine. We aren't trying to build a superintelligence. We're trying to clear our calendars.

Identify the "Cognitive Paperwork"

Drudgery is any task that requires "doing" but not "being." It's the work that doesn't require your unique soul, your specific taste, or your physical presence.

- **The "Blank Page" Problem:** Whether it's writing a polite but firm letter to a contractor, a condolence note, or a volunteer sign-up sheet, starting from zero is hard. Let the AI write the first draft. You shouldn't be the one digging the foundation; you should be the one inspecting the house.
- **The Information Sieve:** We are all drowning in PDFs, long-winded emails, and 60-minute recorded meetings. Use AI to act as your filter. Feed it the transcript or the document and say, "Give me the three most important takeaways and anything that requires me to take action."

The Personal Logistics Officer

The most immediate "win" for a general audience isn't in the office; it's at the kitchen table. AI is the ultimate tool for managing the "admin" of being a human.

- **The Meal-Plan Pivot:** Instead of staring at the fridge in a panic, tell the AI: "I have chicken, half a bag of spinach, and some heavy cream. Give me a recipe that takes less than 20 minutes and doesn't use the oven because it's too hot today."
- **The Comparison Shopper:** Buying a new dishwasher or choosing a summer camp? Don't spend four hours reading conflicting reviews. Ask the AI to "compare these three options based on reliability, price, and noise level, and highlight any common complaints from people who live in high-altitude climates like Colorado."

The Polite Buffer

One of the greatest drains on our energy is the social labor of saying "no" or navigating conflict. AI is a master of tone. If you need to tell a friend you can't make it to their party without sounding like a jerk, or you need to explain to a customer why a project is delayed, let the AI handle the delicate phrasing.

- **Prompt:** "I need to tell my neighbor I can't look after their dog next weekend, but I want to stay on good terms. Make it sound warm but firm."

The goal isn't to let the AI do 100% of the work. The goal is to let it do the first 80%. Let it gather the data, draft the text, and organize the schedule. Then, you step in for the final 20%—the human polish. You add the personal touch, the specific joke, or the final "yes."

The Professional "Filter" Strategy

In a professional setting, the "Drudgery" isn't just boring tasks—it's the "work about work" that prevents you from actually doing the job you were hired for. Most of us spend 60% of our day in a state of shallow work: answering emails, formatting slides, and hunting for information in long message threads.

By applying the delegation strategy to your career, you move from being a "reactive" worker to a "proactive" leader. Here is how to apply the filter:

- **The Meeting Synthesis:** Never walk out of a meeting with three pages of messy notes again. Use a recorder or transcription tool, feed the text to your AI, and ask it to: *"Identify all action items assigned to me, summarize the three biggest disagreements, and draft a follow-up email to the team."* You've turned an hour of administrative cleanup into thirty seconds of review.
- **The "Vibe Check" on Outbound Communication:** We've all spent twenty minutes obsessing over the

tone of a sensitive email to a client or a difficult boss. Delegate the anxiety. Write a "brain dump" of what you want to say—don't worry about being professional—and tell the AI: *"I'm feeling frustrated but I need to sound collaborative and solution-oriented. Fix this email for me."* The machine handles the emotional labor; you just hit send.

- **Data Translation for Non-Experts:** If your job involves looking at spreadsheets or technical reports, use the AI as a translator. Instead of spending hours trying to explain "Quarterly Variance" to a creative team, ask the AI: *"Explain these financial results using a mountain-climbing analogy so it's easy for a group of designers to understand."*

- **The Research Shortcut:** Instead of "Googling" and clicking through ten ads and three pop-up-cluttered articles, use your AI as a research librarian. Ask it to *"Find the current industry standards for creative agency margins in 2026 and summarize the pros and cons of a value-based pricing model."* You aren't avoiding the research; you're skipping the "scrolling" and going straight to the "thinking."

In the workplace, the person who delegates the drudgery isn't the one who is doing less. The delegator is the one contributing more. By offloading the "mechanical" parts of your job to the machine, you free up your brain for the high-value work that only you can do: strategy, creativity, and human connection.

The Bottom Line: Every minute you spend on drudgery is a minute you aren't spending on the things that make life worth living. By delegating the boring stuff to the machine, you aren't being lazy; you are being an architect. You are clearing the digital debris so you can focus on the real construction of your life.

————

Pillar 5: Reclaiming the Analog

If you have followed the first four pillars, something remarkable happens: you start to find "pockets" in your day that weren't there before. By directing instead of doing, learning in five-minute bursts, and delegating the digital drudgery, you have successfully bought back the most precious commodity on Earth: **your time.**

The final, and most important, part of the Architect's Playbook is deciding what to do with it.

There is a subtle trap in the AI era—the temptation to use the time you saved with AI to simply consume more AI. We save twenty minutes on a grocery list only to spend that twenty minutes scrolling through an algorithmically curated feed on our phones. This pillar is about the intentional "U-turn" back to the physical world. It is the realization that the goal of the machine is to get you *off* the screen, not keep you on it.

The "Analog" High Ground

In a world of infinite digital noise, the physical world is the only place left with a "high signal." These are the activities that have no algorithm, no "like" button, and no shortcut. They are the things that ground us in our bodies and our communities.

- **The Ritual of the Real:** Whether it's the physical process of grinding beans and brewing that perfect cup of local coffee, or the tactile feeling of soil in a garden, these rituals are essential for our mental health. They remind us that we are biological creatures, not just data-processing units.

- **The "Shared Air" Connection:** We were designed to be in the same room as other humans. A digital "check-in" is a shadow of a shared meal. Use your reclaimed time to prioritize "Shared Air"—those face-to-face moments with family, friends, or neighbors where the phone is put away and the connection is unfiltered.

Nature as the Ultimate Reset

As an architect of your own life, you must recognize that your brain needs the "randomness" of the natural world to stay creative. The machine is built on logic and patterns. The mountains are built on chaos, weather, and deep time.

- **The "Off-Grid" Recovery:** In Colorado, we have the ultimate antidote to the digital hum right in our backyard. Taking your family into the mountains isn't just a "vacation"—it is a necessary recalibration. It's about moving through a landscape that doesn't care about your notifications or your "output." This "analog reset" is what allows you to return to the digital world with a fresh, human perspective.

The Sovereignty of Focus

The ultimate luxury of the next decade will be the ability to focus on one thing for a long time without interruption. AI can do a thousand things at once, but only a human can experience the deep "flow" of a single pursuit—reading a physical book, practicing an instrument, or having a deep, three-hour conversation.

Reclaiming the analog means protecting your focus. It means

deciding that some parts of your life are "sacred territory" where the machine is simply not invited.

The "Offline" Professional Advantage

In the modern workplace, the ultimate flex is no longer being available on Slack 24/7. It is having the leverage and the systems in place to step away from the screen entirely.

When you use AI to handle the digital drudgery—the meeting summaries, the email drafting, the data sorting—you aren't doing it so you can take on twice as much busywork. You are doing it to clear the deck for the high-impact, offline work that actually moves the needle in a career.

Here is how reclaiming the analog becomes your ultimate professional advantage:

- **The Analog Brainstorm:** AI is unparalleled at synthesizing existing data and generating a hundred variations of an established idea. But true, zero-to-one creative breakthroughs require physical space and human friction. The best campaign strategies and brand visions rarely happen while staring alone at a blinking cursor. They happen when you gather a creative team around a table at a local coffee shop with a notebook or a physical whiteboard. You use the AI *afterward* to organize the mess, but the vital spark happens in the room.
- **Leadership is a Physical Act:** You can automate project management, but you cannot automate mentorship. Whether you are directing a creative studio, serving on an advisory board, or guiding a community campaign, true leadership requires "shared air." Reading the subtle shifts in body language during a pitch, building genuine trust with a client over dinner, and navigating the emotional

nuances of a team are strictly analog skills. The more digital the company becomes, the more valuable the offline leader is.

- **The Strategic Disconnect:** Your brain needs downtime to process complex problems. Staring at a monitor for ten hours a day actively degrades your cognitive function. The professionals who will thrive are the ones who deliberately protect their analog boundaries. Stepping away to sketch out a concept on actual paper, reviewing physical design proofs, or simply walking outside to clear your head provides the necessary friction your brain needs to connect disparate ideas.

Let the machine handle the scale, but protect the time required to handle the soul of the work. The most successful professionals in the AI era will be the ones who master their digital tools, close the laptop, and step back out into the physical world—regularly hunting for the perfect cup of coffee to fuel the next chapter.

The Bottom Line: The "Architect's Playbook" doesn't end with a faster computer or a better prompt. It ends with you standing in the physical world, fully present, and deeply connected to the people and places around you. We use the machine to handle the digital so we can be more human in the analog. The machine is the tool. Your life is the masterpiece.

CONCLUSION

THE ARCHITECT'S ADVANTAGE

We started this journey surrounded by noise. The headlines told us our jobs were ending, our data was being harvested, our power grids were collapsing, and our ultimate creation was quietly preparing to lock us out of the future. We were standing at the edge of the AI era, paralyzed by the panics.

But as we have dismantled these fears one by one, a very different reality has emerged. The arrival of artificial intelligence is not the end of human usefulness. It is the end of human drudgery.

The End of the Human Machine

For the last century, the industrial and corporate worlds convinced us that our highest value was acting like computers. We traded our time, our cognitive energy, and our focus to process data, draft repetitive documents, and execute mechanical workflows. We graded ourselves on how efficiently we could perform tasks that required zero actual soul.

Now, actual machines have finally arrived to take those jobs back.

This leaves us with a profound choice. We can cling to the old ways, fighting a losing, anxious battle against the shifting algorithms of tech giants, or the automated futures mapped out by industrial leaders. Or, we can choose to step up into the director's chair.

The Analog Imperative

The ultimate irony of the artificial intelligence revolution is that it makes our physical humanity more valuable than ever. When flawless digital execution can be synthetically generated in seconds, the messy, uncopyable reality of a human connection becomes the premium currency.

The future does not belong to the people who can write the most code or stare at a screen the longest. It belongs to the architects of life. It belongs to the people who use the machine to handle the digital heavy lifting so they can aggressively reclaim their offline lives. It belongs to those who use their bought-back time to lead their physical communities, hone an analog craft, and actually breathe the air—whether that be the air of excitement in a conference room primed to launch a team on a new assignment or the cool air you take in contemplatively on a mountain trail.

The Blinking Cursor

With AI in hand, perhaps that blinking cursor on the blank page isn't quite as daunting as it used to be. Now you have a confidant to ask, "Where do we go from here?" or "How do we get started?" See it not as the tool that ends the way things used to be, but rather as a tool that helps you bring new visions to life in ways you just couldn't quite do so before.

You now have the playbook. You know how to direct the

intelligence rather than do the task. You know how to keep your skills sharp in five-minute bursts, how to delegate your cognitive paperwork, and how to fiercely protect your "Proof of Human."

The theoretical panic is over. The tools are on the table. The machine is waiting for your instructions.

Close the book.

Put down the device.

Step outside into the real world.

Decide what you want to build next.

ABOUT THE AUTHOR

Glen Day translates the future. At least, he does his earnest best to.

As a creative and content director with global ad agencies, he has spent his career in conversations and collaborations with innovators at tech companies like AWS, Google, and Microsoft, helping them translate their latest advances into everyday human language. He has also had a front-row seat to other areas of innovation like the mobility industry, where he has worked with automotive leaders like Audi, Subaru, and Toyota, as well as behind-the-scenes change drivers like Qualcomm on projects like the Snapdragon Ride automated driving platform.

In addition to his agency work—and occasional book-cover modeling—Glen serves as a board advisor for the Strategic Artificial Intelligence program at the University of Colorado Colorado Springs.

This breadth of access to innovation leaders has given Glen uniquely deep insights into fast-evolving subjects like AI and self-driving cars—along with a clear realization of the massive gap between doomsday hype and physical reality. Glen uses those insights to bring his readers a reality check and an antidote to future-shock.

Based in Colorado, Glen balances the heavy intake of technological research by fiercely protecting his analog life: enjoying the mountains with his family, and regularly tracking down the perfect cup of coffee to fuel the next chapter.

Namaste, y'all.

DON'T PANIC

PUBLISHED BY PROJECTOR MEDIA

Projector House is an independent publishing collective dedicated to clarity. In an era defined by signal noise and manufactured hype, we provide the lens. Through rigorous non-fiction, documentaries, and digital media, we focus on the structural truths of technology, society, and the future. Our mission is simple: to project a clearer picture of the world, one frame at a time.

FOOTNOTES

ACKNOWLEDGEMENTS
1. Adams, Douglas, 1952-2001. The Hitchhiker's Guide to the Galaxy. New York: Harmony Books, 1980.

FORWARD
1. Yuval Noah Harari, *Homo Deus: A Brief History of Tomorrow* (New York: Harper, 2017).
2. Wolfgang Schivelbusch, *"The Railway Journey: The Industrialization of Time and Space in the 19th Century"* Berkeley: University of California Press, 2014, https://static1.1.sqspcdn.com/static/f/221758/18449447/1338242293590/Schivel busch-
3. Elizabeth L. Eisenstein, "The Printing Press as an Agent of Change," *Cambridge University Press*, 1980.

CHAPTER 1
1. Gallup, Saad, Lydia. "More U.S. Workers Fear Technology Making Their Jobs Obsolete." *Gallup*, September 11, 2023. https://news.gallup.com/poll/510551/workers-fear-technology-making-jobs-obsolete.aspx.
2. American Psychological Association, "2023 Work in America Survey: Artificial Intelligence and the Workplace," *American Psychological Association*, September 2023, https://www.apa.org/pubs/reports/work-in-america/2023-work-america-ai-monitoring
3. Joseph Briggs and Devesh Kodnani, "The Potentially Large Effects of Artificial Intelligence on Economic Growth," *Goldman Sachs*, March 26, 2023, https://www.gspublishing.com/content/research/en/reports/2023/03/27/d64e052b-0f6e-45d7-967b-d7be35fabd16.html
4. McKinsey Global Institute, "Jobs Lost, Jobs Gained: Workforce Transitions in a Time of Automation," *McKinsey & Company*, November 28, 2017, https://www.mckinsey.com/~/media/mckinsey/industries/public%20and%20social%20sector/our%20insights/what%20the%20future%20of%20work%20will%20mean%20for%20jobs%20skills%20and%20wages/mgi-jobs-lost-jobs-gained-executive-summary-december-6-2017.pdf
5. Michael Chui et al., "The Economic Potential of Generative AI: The Next Productivity Frontier," *McKinsey & Company*, June 14, 2023, https://www.mckinsey.com/capabilities/tech-and-ai/our-insights/the-economic-potential-of-generative-ai-the-next-productivity-frontier
6. World Economic Forum, "Future of Jobs Report 2023," *World Economic Forum*,

April 30, 2023, https://www.weforum.org/publications/the-future-of-jobs-report-2023

7. World Economic Forum, "Future of Jobs Report 2025," *World Economic Forum,* January 7, 2025 https://www.weforum.org/publications/the-future-of-jobs-report-2025

8. SmartOwner. "THE GREAT DISPLACEMENT | AI JOB LOSS CALCULATOR," n.d. https://thegreatdisplacement.ai

CHAPTER 2

1. Jasper. "Put AI Agents to Work for Marketing | Jasper," n.d. https://www.jasper.ai

2. Journal, Aba. "Latest Version of ChatGPT Aces Bar Exam With Score Nearing 90th Percentile." *ABA Journal,* March 16, 2023. https://www.abajournal.com/web/article/latest-version-of-chatgpt-aces-the-bar-exam-with-score-in-90th-percentile

3. Scott M. McKinney et al., "International evaluation of an AI system for breast cancer screening," *Nature* 577 (January 1, 2020): 89–94, https://pubmed.ncbi.nlm.nih.gov/31894144

4. Eirini Kalliamvakou, "Research: quantifying GitHub Copilot's impact on developer productivity and happiness," *GitHub Blog,* September 7, 2022, https://github.blog/news-insights/research/research-quantifying-github-copilots-impact-on-developer-productivity-and-happiness

5. William Stanley Jevons, "The Coal Question: An Inquiry Concerning the Progress of the Nation, and the Probable Exhaustion of Our Coal-Mines" *London: Macmillan and Co.,* 1865, https://www.connaissancedesenergies.org/sites/connaissancedesenergies.org/files/pdf-actualites/w._stanley_jevons_the_coal_question_1865_.pdf

6. New York City Department of Consumer and Worker Protection, "Automated Employment Decision Tools (AEDT)," *Ogletree Deakins,* April 7, 2023, https://ogletree.com/insights-resources/blog-posts/new-york-city-adopts-final-rules-on-automated-decision-making-tools-ai-in-hiring

CHAPTER 3

1. Heather Chen and Kathleen Magramo, "Finance Worker Pays Out $25 Million After Video Call with Deepfake 'Chief Financial Officer'," *CNN,* February 4, 2024 https://www.cnn.com/2024/02/04/asia/deepfake-cfo-scam-hong-kong-intl-hnk

2. Coalition for Content Provenance and Authenticity (C2PA). "C2PA | Providing Origins of Media Content," n.d. https://c2pa.org

3. Benjamin Weiser, "Here's What Happens When Your Lawyer Uses ChatGPT," *The New York Times,* May 27, 2023, https://www.nytimes.com/2023/05/27/nyregion/avianca-airline-lawsuit-chatgpt.html

4. Gaby Clark, "AI Chatbots Can Run With Medical Misinformation, Study Finds, Highlighting the Need for Stronger Safeguards," *The Mount Sinai Hospital,*

August 06, 2025, https://www.mountsinai.org/about/newsroom/2025/ai-chatbots-can-run-with-medical-misinformation-study-finds-highlighting-the-need-for-stronger-safeguards

5. Michael M. Grynbaum and Ryan Mac, "The Times Sues OpenAI and Microsoft Over A.I. Use of Copyrighted Work," *The New York Times*, December 27, 2023, https://www.nytimes.com/2023/12/27/business/media/new-york-times-open-ai-microsoft-lawsuit.html

6. Blake Brittain, "Major Record Labels Sue AI Music Generators Suno and Udio for Copyright Infringement," *Reuters*, June 24, 2024, https://www.reuters.com/technology/artificial-intelligence/music-labels-sue-ai-companies-suno-udio-us-copyright-infringement-2024-06-24

7. Sara Guaglione. "A Timeline of the Major Deals Between Publishers and AI Tech Companies in 2025." *Digiday*, January 1, 2026. https://digiday.com/media/a-timeline-of-the-major-deals-between-publishers-and-ai-tech-companies-in-2025

CHAPTER 4

1. Sarah Perez, "AI companion apps on track to pull in $120M in 2025," *TechCrunch*, August 12, 2025, https://techcrunch.com/2025/08/12/ai-companion-apps-on-track-to-pull-in-120m-in-2025/

2. Maples, B., Cerit, M., Vishwanath, A. *et al.* Loneliness and suicide mitigation for students using GPT3-enabled chatbots. *npj Mental Health Res* 3, 4 (2024). https://doi.org/10.1038/s44184-023-00047-6

3. Afshin Khadangi, Hanna Marxen, Amir Sartipi, Igor Tchappi, Gilbert Fridgen, "When AI Takes the Couch: Psychometric Jailbreaks Reveal Internal Conflict in Frontier Models," *Cornell University*, 6 Dec 2025, https://arxiv.org/abs/2512.04124

4. Plato. (1901). Phaedrus (B. Jowett, Trans.). In Plato & B. Jowett (Trans.), *Dialogues of Plato: With analyses and introductions,* Vol. 1, pp. 517–585). Charles Scribner's Sons. https://doi.org/10.1037/13728-013

5. Michael Henry Tessler et al., "AI can help humans find common ground in democratic deliberation," *Science* 386, no. 6719, October 17, 2024

CHAPTER 5

1. Donald C. Shoup, *The High Cost of Free Parking*, Chicago: Planners Press, 2005, https://www.researchgate.net/publication/235359727_The_High_Cost_of_Free_Parking

2. European Parliament and Council of the European Union, *Artificial Intelligence Act*, [Date of the specific draft/passage you are referencing, e.g., March 13, 2024], https://www.europarl.europa.eu/doceo/document/TA-9-2024-0138_EN.html

3. Joy Buolamwini, *Unmasking AI: My Mission to Protect What Is Human in a World of Machines* (New York: Random House, 2023)

4. Joy Buolamwini and Timnit Gebru, "Gender Shades: Intersectional Accuracy

Disparities in Commercial Gender Classification," *Proceedings of Machine Learning Research* 81 (2018): 1–15.

5. FIDO Alliance, "Businesses are Ready to Ditch Passwords, Says New Report from FIDO Alliance and LastPass", *FIDO News Center*, October 16, 2023, https://fidoalliance.org/businesses-are-ready-to-ditch-passwords-says-new-report-from-fido-alliance-and-lastpass/

6. U.S. Department of Defense, *Directive 3000.09: Autonomy in Weapon Systems* (Updated January 25, 2023), https://www.esd.whs.mil/portals/54/documents/dd/issuances/dodd/300009p.pdf

7. Thomas Haigh, Mark Priestley, and Crispin Rope, "ENIAC in Action: Making and Remaking the Modern Computer" *Cambridge: MIT Press*, 2016, https://archive.org/details/eniacinactionmak0000thom/page/n5/mode/2up

CHAPTER 6

1. International Energy Agency, *Electricity 2024: Analysis and Forecast to 2026* (Paris: IEA, 2024), https://www.iea.org/reports/electricity-2024

2. U.S. Geological Survey, *"Mineral Commodity Summaries 2024" Reston, VA: U.S. Geological Survey*, 2024, [https://pubs.usgs.gov/publication/mcs2024

3. Amil Merchant et al., "Scaling Deep Learning for Materials Discovery," *Nature* 624, December 2023, P. 80–85, https://www.nature.com/articles/s41586-023-06735-9

4. Alexandra Sasha Luccioni, Yacine Jernite, and Emma Strubell, "Power Hungry Processing: Watts Driving the Cost of AI Deployment?" *arXiv* preprint arXiv:2311.16863 (November 2023): https://arxiv.org/pdf/2311.16863

5. Mark Lynas, Benjamin Z. Houlton, and Simon Perry, "Greater than 99% consensus on human-caused climate change in the peer-reviewed scientific literature," *Environmental Research Letters* 16, no. 11 (2021): 114005,https://ui.adsabs.harvard.edu/abs/2021ERL....16k4005L/abstract

6. Pengfei Li et al., "Making AI Less 'Thirsty': Uncovering and Addressing the Secret Water Footprint of AI Models," *arXiv* preprint arXiv:2304.03271 (April 2023): https://arxiv.org/abs/2304.03271

7. Jake Bittle, "Arizona's water is drying up. That's not stopping the data center rush" *Grist*, [Mar 04, 2026] https://grist.org/technology/arizona-water-data-centers-semiconducters/

8. Luke Barratt and Rosa Furneaux, "Amazon strategised about keeping its data-centres' full water use secret, leaked document shows" *The Guardian*, October 25, 2025, https://www.theguardian.com/technology/2025/oct/25/amazon-datacentres-water-use-disclosure

CHAPTER 7

1. Irving John Good, "Speculations Concerning the First Ultraintelligent Machine," *Science Direct* (1965): https://www.sciencedirect.com/science/chapter/bookseries/abs/pii/S0065245808604180

2. Max Tegmark, *Life 3.0: Being Human in the Age of Artificial Intelligence* (New York: Knopf, 2017).

3. Liv McMahon, "AI system resorts to blackmail if told it will be removed," *BBC*, 23 May 2025, https://www.bbc.com/news/articles/cpqeng9d20go

4. Vittoria Elliott, "AI Steve: The Artificial Intelligence Candidate Running for UK Parliament," *Wired*, Jun 11, 2024, https://www.wired.com/story/ai-candidate-running-for-parliament-uk/

5. Jane C. Timm, "An AI bot for mayor? Wyoming election official says not so fast," *NBC News*, June 17, 2024, https://www.nbcnews.com/politics/2024-election/ai-bot-mayor-cheyenne-wyoming-election-official-rcna157499

6. Sophia Khatsenkova, "Romania's prime minister has hired the world's first AI government adviser. What will it do?" *EuroNews*, June 3, 2023, https://www.euronews.com/next/2023/03/06/romanias-prime-minister-has-hired-the-worlds-first-ai-government-adviser-what-will-it-do

7. François Chollet, "The Implausibility of Intelligence Explosion", *Medium.com*, November 27, 2017, https://medium.com/@francois.chollet/theimpossibility-of-intelligence-explosion-5be4a9eda6ec

8. Reuters, "Musk's Neuralink shows first brain-chip patient playing online chess," *Reuters*, March 20, 2024, https://www.reuters.com/business/healthcare-pharmaceuticals/neuralink-shows-first-brain-chip-patient-playing-online-chess-2024-03-21/

9. Yanis Varoufakis, *Technofeudalism: What Killed Capitalism* (London: The Bodley Head, 2023).

10. Eva Vivalt, Elizabeth Rhodes, Alexander Bartik, David Broockman, Sarah Miller, "The Employment Effects of a Guaranteed Income: Experimental Evidence from Two U.S. States," *Yale University Press*, July 2024, https://economics.yale.edu/sites/default/files/2024-08/Vivalt-et-al.-ORUS-employment.pdf

11. Yuval Noah Harari, *Homo Deus: A Brief History of Tomorrow* (New York: Harper, 2017).

CHAPTER 8

1. Molly Talbert, "How work about work gets in the way of real work," *Asana*, April 17th, 2025, *https://asana.com/resources/why-work-about-work-is-bad*

2. Dafei, Su. "Deep Work Rules for Focused Success in a Distracted World - PDF Room," *IAEME*, 2020, https://www.academia.edu/97189864/Deep_Work_Rules_for_focused_success_in_a_distracted_world_PDF_Room